Speech Enhancement in the Karhunen-Loève Expansion Domain

Synthesis Lectures on Speech and Audio Processing

Editor
B.H. Juang, ***Georgia Tech***

Speech Enhancement in the Karhunen-Loève Expansion Domain
Jacob Benesty, Jingdong Chen, and Yiteng Huang
2011

Sparse Adaptive Filters for Echo Cancellation
Constantin Paleologu, Jacob Benesty, and Silviu Ciochina
June 2010

Multi-Pitch Estimation
Mads Græsbøll Christensen and Andreas Jakobsson
2009

Discriminative Learning for Speech Recognition: Theory and Practice
Xiaodong He and Li Deng
2008

Latent Semantic Mapping: Principles & Applications
Jerome R. Bellegarda
2007

Dynamic Speech Models: Theory, Algorithms, and Applications
Li Deng
2006

Articulation and Intelligibility
Jont B. Allen
2005

Speech Enhancement in the Karhunen-Loève Expansion Domain

Jacob Benesty, Jingdong Chen, and Yiteng Huang

ISBN:978-3-031-01432-1 paperback
ISBN:978-3-031-02560-0 ebook

DOI 10.1007/978-3-031-02560-0

A Publication in the Springer series
SYNTHESIS LECTURES ON SPEECH AND AUDIO PROCESSING
Lecture #7
Series Editor: B.H. Juang, *Georgia Tech*
Series ISSN
Synthesis Lectures on Speech and Audio Processing
Print 1932-121X Electronic 1932-1678

Speech Enhancement in the Karhunen-Loève Expansion Domain

Jacob Benesty
INRS-EMT, University of Quebec

Jingdong Chen
WeVoice, Inc.

Yiteng Huang
WeVoice, Inc.

SYNTHESIS LECTURES ON SPEECH AND AUDIO PROCESSING #7

ABSTRACT

This book is devoted to the study of the problem of speech enhancement whose objective is the recovery of a signal of interest (i.e., speech) from noisy observations. Typically, the recovery process is accomplished by passing the noisy observations through a linear filter (or a linear transformation). Since both the desired speech and undesired noise are filtered at the same time, the most critical issue of speech enhancement resides in how to design a proper optimal filter that can fully take advantage of the difference between the speech and noise statistics to mitigate the noise effect as much as possible while maintaining the speech perception identical to its original form. The optimal filters can be designed either in the time domain or in a transform space. As the title indicates, this book will focus on developing and analyzing optimal filters in the Karhunen-Loève expansion (KLE) domain. We begin by describing the basic problem of speech enhancement and the fundamental principles to solve it in the time domain. We then explain how the problem can be equivalently formulated in the KLE domain. Next, we divide the general problem in the KLE domain into four groups, depending on whether interframe and interband information is accounted for, leading to four linear models for speech enhancement in the KLE domain. For each model, we introduce signal processing measures to quantify the performance of speech enhancement, discuss the formation of different cost functions, and address the optimization of these cost functions for the derivation of different optimal filters. Both theoretical analysis and experiments will be provided to study the performance of these filters and the links between the KLE-domain and time-domain optimal filters will be examined.

KEYWORDS

noise reduction, speech enhancement, single-channel microphone signal processing, Karhunen-Loève expansion (KLE), time domain, KLE domain, Wiener filter, tradeoff filter, maximum signal-to-noise ratio (SNR) filter, minimum variance distortionless response (MVDR) filter.

Contents

CHAPTER 1
Introduction

A signal of interest (usually speech), when picked up by microphones, is inevitably contaminated by unwanted acoustic distortions. Depending on the mechanism that generates them, these distortions can be broadly classified into four basic categories: additive noise originating from various ambient sound sources, interference from concurrent competing speakers, filtering effects caused by room surface reflections and spectral shaping of recording devices, and echo from coupling between loudspeakers and microphones. These four categories of distortions interfere with the measurement, processing, recording, and communication of the desired speech signal in very distinct ways, and combating them has led to four important research areas: speech enhancement (also called noise reduction), source separation, speech dereverberation, and echo cancellation and suppression. A broad coverage of these research areas can be found in [6], [30]. This book is devoted to the study of the problem of single-channel speech enhancement in the Karhunen-Loève expansion (KLE) domain.

1.1 THE PROBLEM OF SPEECH ENHANCEMENT

Speech enhancement consists of recovering a speech signal of interest from microphone observations, which are corrupted by unwanted additive noise. By additive noise, we mean that the signal picked up by a microphone is a superposition of the clean speech and noise. In this scenario, the noise does not directly modify the statistics of the desired speech signal. However, the observed noisy signal can have very different characteristics in comparison to the desired speech. To illustrate this, Fig. 1.1 shows a clean speech signal, the same signal observed in a noisy conference room, and their spectrograms. Inspecting the difference between the clean speech and noisy signal spectrograms, one may notice that the noise effect manifests itself in several different aspects, including but not limited to: 1) many new frequency components are added into the observed signal, 2) a great portion of the time-varying spectra of the desired speech is masked, 3) the spectral intensity is increased, 4) the dynamic properties of the desired speech spectra near phonetic boundaries are smeared, and 5) the intermittent nature of speech becomes less distinct. These changes may greatly affect the human's perception of the desired speech. On the one hand, one can still perceive the useful information embedded in the desired speech signal when listening to the noisy one; but it would take more attention and may easily lead to listening fatigue. On the other hand, it may become impossible to comprehend the desired speech if the noise is strong. As a result, how to mitigate the noise effect, thereby recovering the desired speech signal from its noisy observations, has become an important problem for many applications such as voice communication and human-machine interfaces.

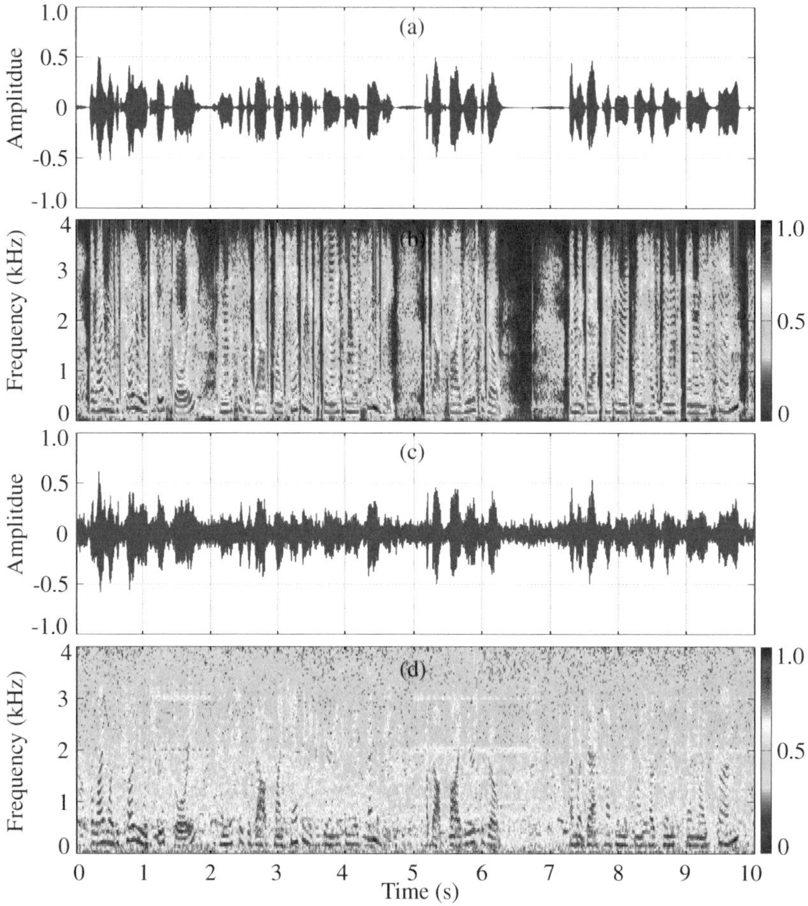

Figure 1.1: Illustration of the noise effect: (a) a clean speech signal, (b) the clean speech spectrogram, (c) a noisy speech observed in a conference room, and (d) the noisy speech spectrogram.

With a use of a single microphone, the noise mitigation process is generally accomplished by properly filtering the noisy speech. The earliest attempt on this was made at Bell Laboratories where Schroeder proposed a system for reducing noise in telecommunication environments in 1960 [48]. His method divides the noisy signal into a number of subbands. For each subband, a rectifier and a lowpass filter are applied in tandem to estimate the noisy speech envelope. The noise level in the corresponding subband is then estimated and subtracted from the noisy speech envelope, resulting in an estimate of the clean speech envelope for the subband. A second rectification process is applied to force the negative results, due to the subtraction, to zero. The rectified clean speech envelope estimate, which is served as a gain filter, is then multiplied with the unmodified subband signal.

Finally, the fullband signal is synthesized from all the subband outputs. This spectral subtraction method implemented with analog circuits, however, has not received much public attention, probably because it was never published in the form of a journal or conference paper for easy and broad circulation.

In the late 1970s, Boll, in his informative paper [9], reformulated the spectral subtraction method but in the framework of digital short-time Fourier analysis, which was later proved to be a particular case of the so-called parametric Wiener filter [41]. Almost at the same time, Lim and Oppenheim, in their landmark work [39], systematically formulated the speech enhancement problem and studied and compared the different algorithms known in the 1970s. Their work demonstrated that speech enhancement was not only effective in improving the quality of noise-corrupted speech, but also useful for increasing both the quality and intelligibility of linear prediction coding (LPC) based parametric speech coding systems. It was this work that had sparkled a huge amount of research attention on the problem. Many algorithms have been developed since then. The most notable contributions include the maximum likelihood (ML) estimator [41], the minimum-mean-square-error (MMSE) estimator [15], [16], and the maximum *a posteriori* (MAP) estimator [51], to name a few. These algorithms share the common key idea of applying a gain (whose value is between 0 and 1) to the noisy speech spectrum in each frequency band to attenuate the noise. They differ only in the form of the gain and how it is estimated.

To derive their gains, the aforementioned MMSE, ML, and MAP estimators assume explicit knowledge of the marginal and joint probability distributions of the clean speech and noise spectra, so that the conditional expected value of the clean speech spectrum, given the noisy speech spectrum, can be evaluated. However, the assumed distributions may not accurately reflect the behavior of the real signals in reality. One way to circumvent this issue is to collect some speech and noise samples and learn the distributions from the collected data. This has led to the development of the hidden Markov model (HMM) based speech enhancement technique. HMM is a statistical model that uses a finite number of states and the associated state transitions to jointly model the temporal and spectral variation of the signals [2]. It has long been used for speech modeling with applications for speech recognition [1], [32], [44]. HMM was introduced to deal with the speech enhancement problem in the late 1980s [17], [18], [19]. This method conquers the problem in two steps. In the first step, which is often called a training process, the probability distributions of the clean speech and the noise process are estimated from given training sequences. The estimated distributions are then applied in the second step to construct speech enhancement filters. Similar to the traditional frequency-domain techniques, the HMM method also applies a gain to the noisy speech spectrum to reduce noise and many different gains can be formed [19], [47]. Besides not requiring an explicit knowledge of the speech and noise distributions, the HMM technique has another advantage of being able to tolerate some nonstationarity in noise, depending on the number of states and mixtures used in the noise HMM. But distortion will arise when the characteristics of the noise are not represented in the training noise data.

Most early attempts in speech enhancement were made in the frequency domain. One may wonder why the frequency domain is preferred to the time domain, given that the noisy signal is originally observed in the time domain and the enhanced signal has to be in the time domain as well. There are many practical reasons for this. First of all, most of our knowledge and understanding of speech production and perception is related to frequencies. In the frequency domain, it is not only easier for us to design speech enhancement filters, but it is more straightforward to analyze and monitor their performance as well. Secondly, thanks to the fast Fourier transform (FFT), the implementation of frequency-domain filters can be made, in general, computationally more efficient than filters in the time domain. Furthermore, the statistics of a speech signal are time and frequency varying and noise can be either white or colored. In the frequency domain, the speech enhancement filters at different frequency bands are designed and handled independently. This gives significant flexibility in exploiting the difference between speech and noise statistics to optimize the amount of noise reduction. However, working in the frequency domain can incur some problems that may not be seen in the time domain, which need special attention. First, due to the circular convolution, some frequency aliasing will be added into the enhanced signal after applying a speech enhancement filter. This problem cannot be completely avoided unless we use a unit gain, which will not give any noise reduction. But one can manage to minimize the effect by applying a proper windowing function (such as the Kaiser one) before FFT and after the inverse FFT (IFFT). Second, speech enhancement filters are generally a function of the noisy and noise spectra. The two spectra are not known *a priori* and have to be estimated in real applications. A *de facto* standard practice in the field of speech enhancement is to treat the short-time FFT spectrum as an estimate of the true spectrum. Such an estimate, however, generally has very large variations about the true spectrum, causing the estimated gains to exceed their theoretical range between 0 and 1. As a result, a nonlinear rectification process has to be used to force the gain to be between 0 and 1. But this would produce some isolated narrowband frequency components in the filtered spectrum. When transformed into the time domain, these isolated components produce music tone sounding noise, which is widely referred to as "musical noise." Musical noise is very unpleasant to hear. Much evidence has shown that listeners would rather prefer to listen to the original noisy signal instead of hearing the enhanced signal with musical noise in most cases. Therefore, it is important not to introduce such noise when we implement a frequency-domain algorithm. But getting rid of musical noise is not a trivial job and it took several decades for engineers to figure out how to do it. Even today, it is still not uncommon to see implementations that result in a signal that is of a lower perceptual quality than the original noisy signal. Because of these problems with the frequency domain techniques, it is often worthwhile to examine the speech enhancement problem in the time domain [4], [7], [11]. The formulation in the time domain not only can avoid some problems with the frequency domain methods, but also can offer new insights into how to design optimal filters and properly evaluate them.

The time and frequency domains are not the only signal spaces in which the speech enhancement problem can be formulated and tackled. In the literature, several other transform spaces have been investigated, such as the LPC model space [22], [23], [35], [37], [42], [43] and the KLE domain. Among them, the KLE domain has received extensive attention. The major difference between the frequency and KLE domains is that the former uses a fixed transform (the Fourier transform) while the latter employs a signal-dependent transform (the KL transform) that is computed from the signal covariance matrix. There are two advantages, at least, of using the signal-dependent KL transform. First, if the covariance matrix is accurately estimated, there will be no aliasing problem. Second, the desired speech and noise may be better separated in the KLE domain than in the frequency domain. The earliest attempts of using the KL transform were made by Dendrinos, Bakamidis, and Carayannis [14] and by Ephraim and Van Trees [20], where the so-called subspace technique was developed. In essence, the subspace approach projects the noisy signal vector into a different domain via the KL transform through the eigenvalue decomposition of an estimate of the correlation matrix of the noisy signal [20]. Once transformed, the speech signal only spans a portion of the entire space, and as a result, the entire vector space can be divided into two subspaces: the signal-plus-noise subspace and the noise-only subspace. Noise reduction is then achieved by removing the noise subspace and cleaning the signal-plus-noise subspace. The rationale of the subspace method in dealing with white noise is rather straightforward; but it becomes less obvious when noise is colored. To cope with colored noise, the subspace approach is extended to a more general form by using the generalized eigenvalue decomposition that simultaneously diagonalizes the clean, noisy, and noise covariance matrices. This extension was first reported in [33] and then redeveloped in [27], [28], [29]. However, one should note that this so-called generalized subspace method is not really a subspace technique since there is no noise-only subspace anymore after the generalized analysis transform. It is more appropriate to call it a constrained Wiener filter. Nevertheless, both the original subspace technique and the generalized one share the common idea of modifying the eigenvalues of the noisy covariance matrix to achieve noise reduction. By analogy, this is similar to filtering the noisy power spectrum in the frequency domain.

Recently, a general formulation of the speech enhancement problem in the KLE domain has been developed [7], [8], [13]. The basic paradigm of this new formulation follows an analysis-filtering-synthesis model. Given a noisy speech signal, which is assumed to be a superposition of a desired clean speech and an unwanted noise signal, a KLE analysis transform will be estimated and applied to transforming a vector of the noisy speech into the KLE domain. Following the convention used in the frequency domain, we call the components corresponding to each KLT base vector a subband. For every subband, a filter is designed and applied to the noisy KLE coefficients, thereby obtaining an estimate of the clean speech KLE coefficients. Finally, the filtered KLE coefficients are transformed back to the time domain using the KLE synthesis transform. The most critical issue with this new formulation is how to design the optimal filters in the KLE domain, which is indeed the focus of this entire book.

1.2 ORGANIZATION OF THE BOOK

The material in this book is organized into nine chapters, including this one. While the focus of the book is on the KLE-domain algorithms as its title indicates, we also attempt to cover the most basic concepts and fundamental principles used to design the optimal filters in the time domain and explain the strong links between the time-domain and KLE-domain filters, which in turn help us better understand how noise reduction works in the frequency domain. The work discussed in these chapters is as follows.

Chapter 2 describes the speech enhancement problem that is going to be dealt with throughout the text. We first formulate the problem in the time domain, and then explain the principles of the KLE and how the time-domain signal model can be equivalently expressed in the KLE domain.

Noisy signals are originally observed in the time domain. It is, therefore, legitimate to tackle the speech enhancement problem in this domain. As pointed earlier, the fundamental issue of speech enhancement in the time domain is how to design a linear filter or a linear transformation that can reduce noise while maintaining the desired speech perception identical to its original form. Typically, the design of a noise reduction filter follows three basic steps: defining a cost function, optimizing the cost function to obtain a noise reduction filter, and evaluating the filter whether it can achieve the expected performance. Chapter 3 provides an overview of the filter design issues in the time domain. We present several performance measures that can be used to evaluate noise reduction filters in the time domain. We also discuss how to define different mean-square errors (MSEs) and how to minimize these MSEs to obtain different noise reduction filters.

In Chapter 4, we discuss the basic speech enhancement problem in the KLE domain and present four linear models depending on whether the interframe and interband information is accounted for. These four linear models will lead to four different filter design approaches in the KLE domain.

Chapters 5 to 8 focus on the optimal noise reduction filter design issues in the KLE domain, with one chapter addressing the design issue associated with one linear model. For each linear model, we discuss the definitions of the performance measures, the MSE cost functions, and how to minimize these cost functions to obtain the optimal noise reduction filters. Also discussed in these chapters are the relationship between the KLE-domain and time-domain filters.

Chapter 9 provides experimental results to validate some of the key filters derived in Chapters 3 and 5–8.

CHAPTER 2

Problem Formulation

In this chapter, we formulate the problem of the additive noise picked up by a microphone along with the desired signal. We also explain the principle of the Karhunen-Loève expansion (KLE) and reformulate the time-domain signal model in the KLE domain.

2.1 SIGNAL MODEL

The noise reduction problem considered in this work is one of recovering the desired signal (or clean speech) $x(k)$, k being the discrete-time index, of zero mean from the noisy observation (microphone signal) [7], [50]

$$y(k) = x(k) + v(k), \tag{2.1}$$

where $v(k)$ is the unwanted additive noise, which is assumed to be a zero-mean random process (white or colored) and uncorrelated with $x(k)$.

The signal model given in (2.1) can be written in a vector form if we process the data by blocks of L samples:

$$\mathbf{y}(m) = \mathbf{x}(m) + \mathbf{v}(m), \tag{2.2}$$

where $m \geq 0$ is the time-frame index,

$$\mathbf{y}(m) = \begin{bmatrix} y(mL) & y(mL+1) & \cdots & y(mL+L-1) \end{bmatrix}^T \tag{2.3}$$

is a vector of length L, superscript ^{T} denotes transposition of a vector or a matrix, and $\mathbf{x}(m)$ and $\mathbf{v}(m)$ are defined in a similar way to $\mathbf{y}(m)$. Since $x(k)$ and $v(k)$ are uncorrelated by assumption, the correlation matrix (of size $L \times L$) of the noisy signal is

$$\begin{aligned} \mathbf{R_y} &= E\left[\mathbf{y}(m)\mathbf{y}^T(m)\right] \\ &= \mathbf{R_x} + \mathbf{R_v}, \end{aligned} \tag{2.4}$$

where $E[\cdot]$ denotes mathematical expectation, and

$$\begin{aligned} \mathbf{R_x} &= E\left[\mathbf{x}(m)\mathbf{x}^T(m)\right], \\ \mathbf{R_v} &= E\left[\mathbf{v}(m)\mathbf{v}^T(m)\right], \end{aligned}$$

are the correlation matrices of $\mathbf{x}(m)$ and $\mathbf{v}(m)$, respectively.

Our objective is then to find a "good" estimate of either $x(k)$ or $\mathbf{x}(m)$ in the sense that the additive noise is significantly reduced while the desired signal is lowly distorted. This book will focus on the estimation of $\mathbf{x}(m)$. For that purpose, we will fully exploit the properties of the KLE.

2.2 KARHUNEN-LOÈVE EXPANSION (KLE)

As explained in [7], [8], [13], it may be advantageous to perform noise reduction in the KLE domain. In this section, we briefly recall the principle of the KLE which can be applied to $\mathbf{y}(m)$, $\mathbf{x}(m)$, or $\mathbf{v}(m)$. In this study, we choose to apply it to $\mathbf{y}(m)$ while the same concept was developed for $\mathbf{x}(m)$ in [7], [8], [13]. Fundamentally, we should not expect much difference between the two, but it is preferable to apply the KLE to $\mathbf{y}(m)$ as the corresponding covariance matrix is usually full rank and well conditioned.

Let us first diagonalize the correlation matrix $\mathbf{R_y}$ as follows [24]:

$$\mathbf{Q}^T \mathbf{R_y} \mathbf{Q} = \mathbf{\Lambda}, \tag{2.5}$$

where

$$\mathbf{Q} = \begin{bmatrix} \mathbf{q}_1 & \mathbf{q}_2 & \cdots & \mathbf{q}_L \end{bmatrix} \tag{2.6}$$

and

$$\mathbf{\Lambda} = \text{diag}\left(\lambda_1, \lambda_2, \ldots, \lambda_L\right) \tag{2.7}$$

are, respectively, orthogonal and diagonal matrices. The orthonormal vectors $\mathbf{q}_1, \mathbf{q}_2, \ldots, \mathbf{q}_L$ are the eigenvectors corresponding, respectively, to the eigenvalues $\lambda_1, \lambda_2, \ldots, \lambda_L$ of the matrix $\mathbf{R_y}$.

The vector $\mathbf{y}(m)$ can be written as a combination (expansion) of the eigenvectors of the correlation matrix $\mathbf{R_y}$ as follows:

$$\mathbf{y}(m) = \sum_{l=1}^{L} c_{y,l}(m)\mathbf{q}_l, \tag{2.8}$$

where

$$c_{y,l}(m) = \mathbf{q}_l^T \mathbf{y}(m), \ l = 1, 2, \ldots, L \tag{2.9}$$

are the coefficients of the expansion and l is the subband[1] index. The representation of the random vector $\mathbf{y}(m)$, described by (2.8) and (2.9), is the Karhunen-Loève expansion (KLE) [25]. Equations (2.8) and (2.9) are, respectively, the synthesis and analysis parts of this expansion.

[1] In this book, the term subband refers to the signal component along each basis vector of the KLE.

From (2.9), we can easily verify that

$$E\left[c_{y,l}(m)\right] = 0,\ l = 1, 2, \ldots, L \tag{2.10}$$

and

$$E\left[c_{y,i}(m)c_{y,j}(m)\right] = \begin{cases} \lambda_i, & i = j \\ 0, & i \neq j \end{cases}. \tag{2.11}$$

It can also be checked from (2.9) that

$$\sum_{l=1}^{L} c_{y,l}^2(m) = \|\mathbf{y}(m)\|_2^2, \tag{2.12}$$

where $\|\mathbf{y}(m)\|_2$ is the Euclidean norm of $\mathbf{y}(m)$. The previous expression shows the energy conservation through the KLE process.

We also define

$$c_{x,l}(m) = \mathbf{q}_l^T \mathbf{x}(m),\ l = 1, 2, \ldots, L, \tag{2.13}$$

$$c_{v,l}(m) = \mathbf{q}_l^T \mathbf{v}(m),\ l = 1, 2, \ldots, L. \tag{2.14}$$

We can check that

$$\sum_{l=1}^{L} c_{x,l}^2(m) = \|\mathbf{x}(m)\|_2^2, \tag{2.15}$$

$$\sum_{l=1}^{L} c_{v,l}^2(m) = \|\mathbf{v}(m)\|_2^2. \tag{2.16}$$

From (2.11), we see that the interband correlation of the coefficients $c_{y,l}(m)$ is equal to 0. But the interband correlations of the coefficients $c_{x,l}(m)$ and $c_{v,l}(m)$ are

$$E\left[c_{x,i}(m)c_{x,j}(m)\right] = \mathbf{q}_i^T \mathbf{R_x} \mathbf{q}_j, \tag{2.17}$$

$$E\left[c_{v,i}(m)c_{v,j}(m)\right] = \mathbf{q}_i^T \mathbf{R_v} \mathbf{q}_j. \tag{2.18}$$

It is easy to verify that these interband correlations, i.e., $E\left[c_{x,i}(m)c_{x,j}(m)\right]$ and $E\left[c_{v,i}(m)c_{v,j}(m)\right]$ for $i \neq j$, are equal to 0 only when the noise is white (assuming that the desired signal, i.e., speech, is always correlated which is usually the case). However, in practice, noise is rarely white and the interband correlation should be taken into account in the design of filters for noise reduction. This idea was first proposed in [38] but in the frequency domain.

The speech signal is highly correlated. Therefore, the interframe correlation cannot be expected to be zero, i.e., $E\left[c_{x,l}(m)c_{x,l}(m-i)\right] \neq 0$, and should be considered in the development of noise reduction algorithms.

Left multiplying both sides of (2.2) by $\mathbf{q}_l^T$, the time-domain signal model is transformed into the KLE domain as

$$c_{y,l}(m) = c_{x,l}(m) + c_{v,l}(m), \; l = 1, 2, \ldots, L. \tag{2.19}$$

Therefore, noise reduction in the KLE domain corresponds to the estimation of the coefficients $c_{x,l}(m), \; l = 1, 2, \ldots, L$, from the observations $c_{y,l}(m), \; l = 1, 2, \ldots, L$ [7], [8], [13].

CHAPTER 3

Optimal Filters in the Time Domain

This chapter reviews the classical time-domain linear filtering technique for noise reduction. Some new results are also presented as well. This chapter is important for the rest of this work since we will show later some interesting and strong links with noise reduction in the KLE domain.

In the time domain, the objective of noise reduction is to estimate $\mathbf{x}(m)$ from the observation vector $\mathbf{y}(m)$. Usually, we estimate the noise-free speech, $\mathbf{x}(m)$, by applying a linear transformation to the microphone signal [4], [5], [12], [30], [40], [50], i.e.,

$$\begin{aligned} \mathbf{z}(m) &= \mathbf{H}_{\mathrm{t}}\mathbf{y}(m) \\ &= \mathbf{H}_{\mathrm{t}}\left[\mathbf{x}(m) + \mathbf{v}(m)\right] \\ &= \mathbf{x}_{\mathrm{f}}(m) + \mathbf{v}_{\mathrm{rn}}(m), \end{aligned} \tag{3.1}$$

where $\mathbf{H}_{\mathrm{t}}$ is a filtering matrix of size $L \times L$,

$$\mathbf{x}_{\mathrm{f}}(m) = \mathbf{H}_{\mathrm{t}}\mathbf{x}(m) \tag{3.2}$$

is the filtered clean speech (or filtered desired signal), and

$$\mathbf{v}_{\mathrm{rn}}(m) = \mathbf{H}_{\mathrm{t}}\mathbf{v}(m) \tag{3.3}$$

is the filtered noise, which is often called the residual noise. The correlation matrix of the estimated signal is then

$$\begin{aligned} \mathbf{R}_{\mathbf{z}} &= E\left[\mathbf{z}(m)\mathbf{z}^T(m)\right] \\ &= \mathbf{H}_{\mathrm{t}}\mathbf{R}_{\mathbf{x}}\mathbf{H}_{\mathrm{t}}^T + \mathbf{H}_{\mathrm{t}}\mathbf{R}_{\mathbf{v}}\mathbf{H}_{\mathrm{t}}^T. \end{aligned} \tag{3.4}$$

Therefore, with this time-domain formulation, the noise reduction problem becomes one of finding "good" filtering matrices that would attenuate the noise as much as possible while keeping the clean speech from being dramatically distorted.

We start this chapter by defining some important performance measures.

3.1 PERFORMANCE MEASURES

One of the most important measures in noise reduction is the signal-to-noise ratio (SNR). We define the input SNR as the ratio of the intensity of the signal of interest (speech) over the intensity of the background noise, i.e.,

$$\mathrm{iSNR} = \frac{\sigma_x^2}{\sigma_v^2}, \tag{3.5}$$

where

$$\sigma_x^2 = E\left[x^2(k)\right]$$

and

$$\sigma_v^2 = E\left[v^2(k)\right]$$

are the variances of the signals $x(k)$ and $v(k)$, respectively. This definition of the input SNR can also be written in another form. With the signal model shown in (2.2), it is easy to check that

$$\sigma_x^2 = \frac{\mathrm{tr}\left(\mathbf{R_x}\right)}{L}$$

and

$$\sigma_v^2 = \frac{\mathrm{tr}\left(\mathbf{R_v}\right)}{L},$$

where $\mathrm{tr}(\cdot)$ denotes the trace of a square matrix. Therefore, the input SNR can be rewritten as

$$\mathrm{iSNR} = \frac{\mathrm{tr}\left(\mathbf{R_x}\right)}{\mathrm{tr}\left(\mathbf{R_v}\right)}. \tag{3.6}$$

After noise reduction with the time-domain model given in (3.1), the output SNR can be expressed as

$$\begin{aligned} \mathrm{oSNR}(\mathbf{H}_\mathrm{t}) &= \frac{E\left[\mathbf{x}_\mathrm{f}^T(m)\mathbf{x}_\mathrm{f}(m)\right]}{E\left[\mathbf{v}_\mathrm{rn}^T(m)\mathbf{v}_\mathrm{rn}(m)\right]} \\ &= \frac{\mathrm{tr}\left(\mathbf{H}_\mathrm{t}\mathbf{R_x}\mathbf{H}_\mathrm{t}^T\right)}{\mathrm{tr}\left(\mathbf{H}_\mathrm{t}\mathbf{R_v}\mathbf{H}_\mathrm{t}^T\right)}. \end{aligned} \tag{3.7}$$

One of the most important goals of noise reduction is to improve the SNR after filtering [5], [11]. Therefore, we must design a filter, $\mathbf{H}_\mathrm{t}$, in such a way that $\mathrm{oSNR}(\mathbf{H}_\mathrm{t}) \geq \mathrm{iSNR}$.

Another important measure in noise reduction is the noise-reduction factor, which quantifies the amount of noise being attenuated by the filter. With the time-domain formulation, this factor is defined as [5], [11]

$$\xi_\mathrm{nr}\left(\mathbf{H}_\mathrm{t}\right) = \frac{\mathrm{tr}\left(\mathbf{R_v}\right)}{\mathrm{tr}\left(\mathbf{H}_\mathrm{t}\mathbf{R_v}\mathbf{H}_\mathrm{t}^T\right)}. \tag{3.8}$$

The larger the value of $\xi_{\mathrm{nr}}(\mathbf{H}_{\mathrm{t}})$, the more the noise is reduced. After the filtering operation, the residual noise level is expected to be lower than that of the original noise level; therefore, this factor should have a lower bound of 1 for optimal filters.

The filtering operation adds distortion to the speech signal. In order to evaluate the amount of speech distortion, the concept of speech-distortion index has been introduced in [5], [11]. With this time-domain model, the speech-distortion index is defined as

$$\begin{aligned} \upsilon_{\mathrm{sd}}(\mathbf{H}_{\mathrm{t}}) &= \frac{E\left\{[\mathbf{x}_{\mathrm{f}}(m)-\mathbf{x}(m)]^T[\mathbf{x}_{\mathrm{f}}(m)-\mathbf{x}(m)]\right\}}{E\left[\mathbf{x}^T(m)\mathbf{x}(m)\right]} \\ &= \frac{E\left\{\left[\mathbf{H}_{\mathrm{t}}\mathbf{x}(m)-\mathbf{x}(m)\right]^T\left[\mathbf{H}_{\mathrm{t}}\mathbf{x}(m)-\mathbf{x}(m)\right]\right\}}{\mathrm{tr}\left(\mathbf{R}_{\mathbf{x}}\right)} \\ &= \frac{\mathrm{tr}\left[(\mathbf{H}_{\mathrm{t}}-\mathbf{I})\mathbf{R}_{\mathbf{x}}(\mathbf{H}_{\mathrm{t}}-\mathbf{I})^T\right]}{\mathrm{tr}\left(\mathbf{R}_{\mathbf{x}}\right)}, \end{aligned} \tag{3.9}$$

where $\mathbf{I}$ is the identity matrix of size $L \times L$. The speech-distortion index has a lower bound of 0 and an upper bound of 1 for optimal filters. The higher the value of $\upsilon_{\mathrm{sd}}(\mathbf{H}_{\mathrm{t}})$, the more the speech is distorted.

A measure that is somewhat similar to the noise-reduction factor is the speech-reduction factor defined as [7]

$$\xi_{\mathrm{sr}}(\mathbf{H}_{\mathrm{t}}) = \frac{\mathrm{tr}\left(\mathbf{R}_{\mathbf{x}}\right)}{\mathrm{tr}\left(\mathbf{H}_{\mathrm{t}}\mathbf{R}_{\mathbf{x}}\mathbf{H}_{\mathrm{t}}^T\right)}. \tag{3.10}$$

The larger the value of $\xi_{\mathrm{sr}}(\mathbf{H}_{\mathrm{t}})$, the more the speech is reduced (or distorted). After the filtering operation, the speech level is typically lower than that of the original speech level; therefore, this factor should have a lower bound of 1 for optimal filters.

It is easy to verify that we always have

$$\frac{\mathrm{oSNR}(\mathbf{H}_{\mathrm{t}})}{\mathrm{iSNR}} = \frac{\xi_{\mathrm{nr}}(\mathbf{H}_{\mathrm{t}})}{\xi_{\mathrm{sr}}(\mathbf{H}_{\mathrm{t}})}. \tag{3.11}$$

3.2 MEAN-SQUARE ERROR (MSE) CRITERION

Although many different criteria can be defined, the mean-square error (MSE) is, by far, the most used one because of its simplicity in terms of deriving useful filters and closed-form estimators.

We define the error signal vector between the estimated and desired signals as

$$\begin{aligned} \mathbf{e}(m) &= \mathbf{z}(m) - \mathbf{x}(m) \\ &= \mathbf{H}_{\mathrm{t}}\mathbf{y}(m) - \mathbf{x}(m), \end{aligned} \tag{3.12}$$

which can also be written as the sum of two orthogonal error signal vectors:

$$\mathbf{e}(m) = \mathbf{e}_x(m) + \mathbf{e}_v(m), \tag{3.13}$$

where

$$\mathbf{e}_x(m) = (\mathbf{H}_\mathrm{t} - \mathbf{I})\,\mathbf{x}(m) \tag{3.14}$$

is the speech distortion due to the linear transformation and

$$\mathbf{e}_v(m) = \mathbf{H}_\mathrm{t}\mathbf{v}(m) \tag{3.15}$$

represents the residual noise [20].

Having defined the error signal, we can now write the MSE criterion:

$$\begin{aligned} J(\mathbf{H}_\mathrm{t}) &= \mathrm{tr}\left\{E\left[\mathbf{e}(m)\mathbf{e}^T(m)\right]\right\} \\ &= \mathrm{tr}(\mathbf{R}_\mathbf{x}) + \mathrm{tr}\left(\mathbf{H}_\mathrm{t}\mathbf{R}_\mathbf{y}\mathbf{H}_\mathrm{t}^T\right) - 2\mathrm{tr}\left(\mathbf{H}_\mathrm{t}\mathbf{R}_\mathbf{yx}\right) \\ &= \mathrm{tr}(\mathbf{R}_\mathbf{x}) + \mathrm{tr}\left(\mathbf{H}_\mathrm{t}\mathbf{R}_\mathbf{y}\mathbf{H}_\mathrm{t}^T\right) - 2\mathrm{tr}\left(\mathbf{H}_\mathrm{t}\mathbf{R}_\mathbf{x}\right), \end{aligned} \tag{3.16}$$

where

$$\mathbf{R}_\mathbf{yx} = E\left[\mathbf{y}(m)\mathbf{x}^T(m)\right]$$

is the cross-correlation matrix between the observation and desired signals, which can also be expressed as

$$\mathbf{R}_\mathbf{yx} = \mathbf{R}_\mathbf{x}$$

since $\mathbf{R}_\mathbf{vx} = E\left[\mathbf{v}(m)\mathbf{x}^T(m)\right] = \mathbf{0}$ [$x(m)$ and $v(m)$ are assumed to be uncorrelated]. Similarly, using the uncorrelation assumption, expression (3.16) can be structured in terms of two MSEs, i.e.,

$$\begin{aligned} J(\mathbf{H}_\mathrm{t}) &= \mathrm{tr}\left\{E\left[\mathbf{e}_x(m)\mathbf{e}_x^T(m)\right]\right\} + \mathrm{tr}\left\{E\left[\mathbf{e}_v(m)\mathbf{e}_v^T(m)\right]\right\} \\ &= J_x(\mathbf{H}_\mathrm{t}) + J_v(\mathbf{H}_\mathrm{t}). \end{aligned} \tag{3.17}$$

For the particular transformation $\mathbf{H}_\mathrm{t} = \mathbf{I}$ (the identity matrix), we get

$$J(\mathbf{I}) = \mathrm{tr}(\mathbf{R}_\mathbf{v}), \tag{3.18}$$

so there will be neither noise reduction nor speech distortion. Using this particular case of the MSE, we define the normalized MSE (NMSE) as

$$\begin{aligned} \tilde{J}(\mathbf{H}_\mathrm{t}) &= \frac{J(\mathbf{H}_\mathrm{t})}{J(\mathbf{I})} \\ &= \mathrm{iSNR} \cdot \upsilon_\mathrm{sd}(\mathbf{H}_\mathrm{t}) + \frac{1}{\xi_\mathrm{nr}(\mathbf{H}_\mathrm{t})}, \end{aligned} \tag{3.19}$$

where

$$\upsilon_{\mathrm{sd}}\left(\mathbf{H}_{\mathrm{t}}\right) = \frac{J_x\left(\mathbf{H}_{\mathrm{t}}\right)}{\mathrm{tr}\left(\mathbf{R_x}\right)}, \tag{3.20}$$

$$\xi_{\mathrm{nr}}\left(\mathbf{H}_{\mathrm{t}}\right) = \frac{\mathrm{tr}\left(\mathbf{R_v}\right)}{J_v\left(\mathbf{H}_{\mathrm{t}}\right)}. \tag{3.21}$$

This shows the connection between the NMSE and the speech-distortion index and the noise-reduction factor defined in Section 3.1.

3.3 WIENER FILTER

If we differentiate the MSE criterion, $J\left(\mathbf{H}_{\mathrm{t}}\right)$ [eq. (3.16)], with respect to $\mathbf{H}_{\mathrm{t}}$ and equate the result to zero, we easily find the Wiener filtering matrix:

$$\begin{aligned}\mathbf{H}_{\mathrm{t,W}} &= \mathbf{R_x R_y^{-1}} \\ &= \mathbf{I} - \mathbf{R_v R_y^{-1}}.\end{aligned} \tag{3.22}$$

This optimal filtering matrix depends on the correlation matrices $\mathbf{R_y}$ and $\mathbf{R_v}$: the first one can be estimated during speech-and-noise periods while the second one can be estimated during noise-only intervals, assuming that the statistics of the noise do not change much with time.

Now, if we substitute (2.5) into (3.22), we get another useful form of the time-domain Wiener filtering matrix:

$$\mathbf{H}_{\mathrm{t,W}} = \mathbf{Q}\left[\boldsymbol{\Lambda} - \mathbf{Q}^T\mathbf{R_v}\mathbf{Q}\right]\boldsymbol{\Lambda}^{-1}\mathbf{Q}^T. \tag{3.23}$$

Let us define the following normalized correlation matrices:

$$\tilde{\mathbf{R}}_{\mathbf{v}} = \frac{\mathbf{R_v}}{\sigma_v^2},$$

$$\tilde{\mathbf{R}}_{\mathbf{x}} = \frac{\mathbf{R_x}}{\sigma_x^2}.$$

A third way to write Wiener is

$$\mathbf{H}_{\mathrm{t,W}} = \tilde{\mathbf{R}}_{\mathbf{x}}\tilde{\mathbf{R}}_{\mathbf{v}}^{-1}\left[\frac{\mathbf{I}}{\mathrm{iSNR}} + \tilde{\mathbf{R}}_{\mathbf{x}}\tilde{\mathbf{R}}_{\mathbf{v}}^{-1}\right]^{-1}. \tag{3.24}$$

We can see from (3.24) that

$$\lim_{\mathrm{iSNR}\to\infty}\mathbf{H}_{\mathrm{t,W}} = \mathbf{I}, \tag{3.25}$$

$$\lim_{\mathrm{iSNR}\to 0}\mathbf{H}_{\mathrm{t,W}} = \mathbf{0}. \tag{3.26}$$

Clearly, the Wiener filtering matrix may have a disastrous effect for low input SNRs since it may remove everything (noise and speech).

Property 3.1 With the optimal Wiener filtering matrix given in (3.22), the output SNR is always greater than or equal to the input SNR, i.e., $\text{oSNR}(\mathbf{H}_{\text{t,W}}) \geq \text{iSNR}$.

Proof. See [7]. □

The minimum MSE (MMSE) and minimum NMSE (MNMSE) are obtained by replacing $\mathbf{H}_{\text{t,W}}$ in (3.16) and (3.19):

$$\begin{aligned} J\left(\mathbf{H}_{\text{t,W}}\right) &= \text{tr}\left(\mathbf{R}_{\mathbf{x}}\right) - \text{tr}\left(\mathbf{R}_{\mathbf{x}}\mathbf{R}_{\mathbf{y}}^{-1}\mathbf{R}_{\mathbf{x}}\right) \\ &= \text{tr}\left(\mathbf{R}_{\mathbf{v}}\right) - \text{tr}\left(\mathbf{R}_{\mathbf{v}}\mathbf{R}_{\mathbf{y}}^{-1}\mathbf{R}_{\mathbf{v}}\right), \end{aligned} \tag{3.27}$$

$$\tilde{J}\left(\mathbf{H}_{\text{t,W}}\right) = 1 - \frac{\text{tr}\left(\mathbf{R}_{\mathbf{v}}\mathbf{R}_{\mathbf{y}}^{-1}\mathbf{R}_{\mathbf{v}}\right)}{\text{tr}\left(\mathbf{R}_{\mathbf{v}}\right)} \leq 1. \tag{3.28}$$

We can compute the speech-distortion index by substituting (3.22) into (3.9):

$$\upsilon_{\text{sd}}\left(\mathbf{H}_{\text{t,W}}\right) = 1 - \frac{\text{oSNR}(\mathbf{H}_{\text{t,W}}) + 2}{\text{iSNR} \cdot \xi_{\text{nr}}\left(\mathbf{H}_{\text{t,W}}\right)} \leq 1. \tag{3.29}$$

Using (3.19) and (3.29), we get the noise-reduction factor:

$$\xi_{\text{nr}}\left(\mathbf{H}_{\text{t,W}}\right) = \frac{\text{oSNR}(\mathbf{H}_{\text{t,W}}) + 1}{\text{iSNR} - \tilde{J}\left(\mathbf{H}_{\text{t,W}}\right)} \geq 1. \tag{3.30}$$

Property 3.2 We have

$$\frac{\text{iSNR}}{1 + \text{oSNR}(\mathbf{H}_{\text{t,W}})} \leq \tilde{J}\left(\mathbf{H}_{\text{t,W}}\right) \leq \frac{\text{iSNR}}{1 + \text{iSNR}}, \tag{3.31}$$

$$\frac{\left[1 + \text{oSNR}(\mathbf{H}_{\text{t,W}})\right]^2}{\text{iSNR} \cdot \text{oSNR}(\mathbf{H}_{\text{t,W}})} \leq \xi_{\text{nr}}\left(\mathbf{H}_{\text{t,W}}\right) \leq \frac{(1 + \text{iSNR})\left[1 + \text{oSNR}(\mathbf{H}_{\text{t,W}})\right]}{\text{iSNR}^2}, \tag{3.32}$$

$$\frac{1}{\left[1 + \text{oSNR}(\mathbf{H}_{\text{t,W}})\right]^2} \leq \upsilon_{\text{sd}}\left(\mathbf{H}_{\text{t,W}}\right) \leq \frac{1 + \text{oSNR}(\mathbf{H}_{\text{t,W}}) - \text{iSNR}}{(1 + \text{iSNR})\left[1 + \text{oSNR}(\mathbf{H}_{\text{t,W}})\right]}. \tag{3.33}$$

Proof. See [7]. □

PARTICULAR CASE: WHITE NOISE

We assume here that the noise picked up by the microphone is white (i.e., $\mathbf{R_v} = \sigma_v^2 \mathbf{I}$). In this situation, the Wiener filtering matrix becomes

$$\mathbf{H}_{\mathrm{t,W}} = \mathbf{I} - \sigma_v^2 \mathbf{R}_{\mathbf{y}}^{-1}, \tag{3.34}$$

where

$$\mathbf{R_y} = \mathbf{R_x} + \sigma_v^2 \mathbf{I}.$$

It is well known that the inverse of the Toeplitz matrix $\mathbf{R_y}$ can be factorized as follows [3], [34]:

$$\mathbf{R}_{\mathbf{y}}^{-1} = \begin{bmatrix} 1 & -c_{21} & \cdots & -c_{L1} \\ -c_{12} & 1 & \cdots & -c_{L2} \\ \vdots & \vdots & \ddots & \vdots \\ -c_{1L} & -c_{2L} & \cdots & 1 \end{bmatrix} \begin{bmatrix} 1/E_1 & 0 & \cdots & 0 \\ 0 & 1/E_2 & \cdots & 0 \\ \vdots & \vdots & \ddots & \vdots \\ 0 & 0 & \cdots & 1/E_L \end{bmatrix}, \tag{3.35}$$

where the columns of the first matrix on the right-hand side of (3.35) are the linear interpolators of the signal $y(k)$ and the elements E_l in the diagonal matrix are the respective interpolation-error powers.

Using the factorization of $\mathbf{R}_{\mathbf{y}}^{-1}$ in (3.27) and (3.28), the MMSE and MNMSE can be rewritten, respectively, as

$$J(\mathbf{H}_{\mathrm{t,W}}) = L\sigma_v^2 - \left(\sigma_v^2\right)^2 \sum_{l=1}^{L} \frac{1}{E_l}, \tag{3.36}$$

$$\tilde{J}(\mathbf{H}_{\mathrm{t,W}}) = 1 - \frac{\sigma_v^2}{L} \sum_{l=1}^{L} \frac{1}{E_l}. \tag{3.37}$$

Assume that the noise-free speech signal, $x(k)$, is very well predictable. In this scenario, $E_l \approx \sigma_v^2,\ \forall\, l$, and replacing this value in (3.37), we find that $\tilde{J}(\mathbf{H}_{\mathrm{t,W}}) \approx 0$. From (3.19), we then deduce that $\upsilon_{\mathrm{sd}}(\mathbf{H}_{\mathrm{t,W}}) \approx 0$ (no speech distortion) and $\xi_{\mathrm{nr}}(\mathbf{H}_{\mathrm{t,W}}) \approx \infty$ (infinite noise reduction). Notice that, from a theoretical point of view (and with white noise), this result is independent of the SNR. Also,

$$\mathbf{H}_{\mathrm{t,W}} \approx \begin{bmatrix} 0 & c_{12} & \cdots & c_{1L} \\ c_{21} & 0 & \cdots & c_{2L} \\ \vdots & \vdots & \ddots & \vdots \\ c_{L1} & c_{L2} & \cdots & 0 \end{bmatrix} \tag{3.38}$$

and $\mathbf{H}_{\mathrm{t,W}}\mathbf{x}(m) \approx \mathbf{x}(m)$, so that $\xi_{\mathrm{sr}}(\mathbf{H}_{\mathrm{t,W}}) \approx 1$ and $\mathrm{oSNR}(\mathbf{H}_{\mathrm{t,W}}) \approx \infty$; therefore, we can almost perfectly recover the signal $x(k)$.

At the other extreme case, let us see now what happens when the signal of interest $x(k)$ is not predictable at all. In this situation, $E_l \approx \sigma_y^2, \ \forall\, l$ and $c_{ij} \approx 0, \ \forall\, i, j, \ i \neq j$. Using these values, we get

$$\mathbf{H}_{\mathrm{t,W}} \approx \frac{\mathrm{iSNR}}{1+\mathrm{iSNR}}\mathbf{I}, \tag{3.39}$$

$$\tilde{J}(\mathbf{H}_{\mathrm{t,W}}) \approx \frac{\mathrm{iSNR}}{1+\mathrm{iSNR}}. \tag{3.40}$$

With the help of the two previous equations, it is straightforward to obtain

$$\xi_{\mathrm{nr}}(\mathbf{H}_{\mathrm{t,W}}) \approx \left(1+\frac{1}{\mathrm{iSNR}}\right)^2, \tag{3.41}$$

$$\upsilon_{\mathrm{sd}}(\mathbf{H}_{\mathrm{t,W}}) \approx \frac{1}{(1+\mathrm{iSNR})^2}, \tag{3.42}$$

$$\mathrm{SNR}(\mathbf{H}_{\mathrm{t,W}}) \approx \mathrm{iSNR}. \tag{3.43}$$

While some noise reduction is achieved (at the price of speech distortion), there is no improvement in the SNR, meaning that the Wiener filter has no positive effect on the microphone signal $y(k)$.

This analysis, even though simple, is quite insightful. It shows that the Wiener filter can mitigate the noise effect and improve the SNR, as long as the desired signal is somewhat predictable. However, in practice some discontinuities could be heard from a voiced signal to an unvoiced one, since for the former the noise will be mostly removed while it will not for the latter.

3.4 TRADEOFF FILTERS

The time-domain NMSE as shown in (3.19) is the sum of two terms. One depends on the speech distortion while the other one depends on the noise reduction. Instead of minimizing the NMSE with respect to $\mathbf{H}_{\mathrm{t}}$ as we already did to find the Wiener filter, we can minimize the speech-distortion index with the constraint that the noise-reduction factor is equal to a value that is greater than one. Mathematically, this is equivalent to

$$\min_{\mathbf{H}_{\mathrm{t}}} J_x(\mathbf{H}_{\mathrm{t}}) \quad \text{subject to} \quad J_v(\mathbf{H}_{\mathrm{t}}) = \beta \cdot \mathrm{tr}\left(\mathbf{R_v}\right), \tag{3.44}$$

where $0 < \beta < 1$ in order to have some noise reduction. If we use a Lagrange multiplier, μ, to adjoin the constraint to the cost function, (3.44) can be rewritten as

$$\mathbf{H}_{\mathrm{t,T},\mu} = \arg\min_{\mathbf{H}_{\mathrm{t}}} \mathcal{L}(\mathbf{H}_{\mathrm{t}}, \mu), \tag{3.45}$$

with

$$\mathcal{L}(\mathbf{H}_{\mathrm{t}}, \mu) = J_x\left(\mathbf{H}_{\mathrm{t}}\right) + \mu\left[J_v\left(\mathbf{H}_{\mathrm{t}}\right) - \beta \cdot \mathrm{tr}\left(\mathbf{R_v}\right)\right] \tag{3.46}$$

and $\mu \geq 0$. From (3.45) and assuming that the sum matrix $\mathbf{R_x} + \mu \mathbf{R_v}$ is invertible (if it is not, the pseudo inverse can be used), we can easily derive the optimal filtering matrix:

$$\begin{aligned}
\mathbf{H}_{\mathrm{t,T},\mu} &= \mathbf{R_x} \left(\mathbf{R_x} + \mu \mathbf{R_v}\right)^{-1} \\
&= \left(\mathbf{R_y} - \mathbf{R_v}\right) \left[\mathbf{R_y} + (\mu - 1)\mathbf{R_v}\right]^{-1} \\
&= \left[(1-\mu)\mathbf{I} + \mu \mathbf{H}_{\mathrm{t,W}}^{-1}\right]^{-1},
\end{aligned} \tag{3.47}$$

where the Lagrange multiplier, μ, satisfies $J_v\left(\mathbf{H}_{\mathrm{t,T},\mu}\right) = \beta \cdot \mathrm{tr}\left(\mathbf{R_v}\right)$, which implies that

$$\xi_{\mathrm{nr}}(\mathbf{H}_{\mathrm{t,T},\mu}) = \frac{1}{\beta} > 1. \tag{3.48}$$

In practice, it is not easy to determine the optimal μ. Therefore, when this parameter is chosen in an ad-hoc way, we can see that for

- $\mu = 1$, $\mathbf{H}_{\mathrm{t,T},1} = \mathbf{H}_{\mathrm{t,W}}$, so the tradeoff filter degenerates to the Wiener one;
- $\mu = 0$, $\mathbf{H}_{\mathrm{t,T},0} = \mathbf{I}$, which is an identity filtering matrix that passes the noisy speech without changing it;
- $\mu > 1$, results in low residual noise at the expense of high speech distortion;
- $\mu < 1$, leads to little speech distortion and little noise reduction.

Property 3.3 With the tradeoff filtering matrix given in (3.47), the output SNR is always greater than or equal to the input SNR, i.e., $\mathrm{oSNR}(\mathbf{H}_{\mathrm{t,T},\mu}) \geq \mathrm{iSNR}, \ \forall \mu \geq 0$.

Proof. See [7]. □

We can find another tradeoff filtering matrix by minimizing the residual noise with the constraint that some level of speech distortion is allowed. Mathematically, this is equivalent to

$$\min_{\mathbf{H}_{\mathrm{t}}} J_v(\mathbf{H}_{\mathrm{t}}) \quad \text{subject to} \quad J_x(\mathbf{H}_{\mathrm{t}}) = \beta_2 \cdot \mathrm{tr}\left(\mathbf{R_x}\right), \tag{3.49}$$

where $\beta_2 > 0$ in order to have some noise reduction. If we use a Lagrange multiplier, μ_2, to adjoin the constraint to the cost function, (3.49) can be rewritten as

$$\mathbf{H}_{\mathrm{t,T},2,\mu_2} = \arg\min_{\mathbf{H}_{\mathrm{t}}} \mathcal{L}(\mathbf{H}_{\mathrm{t}}, \mu_2), \tag{3.50}$$

with

$$\mathcal{L}(\mathbf{H}_{\mathrm{t}}, \mu_2) = J_v\left(\mathbf{H}_{\mathrm{t}}\right) + \mu_2 \left[J_x\left(\mathbf{H}_{\mathrm{t}}\right) - \beta_2 \cdot \mathrm{tr}\left(\mathbf{R_x}\right)\right] \tag{3.51}$$

and $\mu_2 > 0$. The optimal solution to this optimization problem is

$$\mathbf{H}_{\mathrm{t,T},2,\mu_2} = \mathbf{R_x}\left(\mathbf{R_x} + \frac{\mathbf{R_v}}{\mu_2}\right)^{-1}, \tag{3.52}$$

where the Lagrange multiplier, μ_2, satisfies $J_x\left(\mathbf{H}_{\mathrm{t,T},2,\mu_2}\right) = \beta_2 \cdot \mathrm{tr}\left(\mathbf{R_x}\right)$, which implies that

$$\upsilon_{\mathrm{sd}}(\mathbf{H}_{\mathrm{t,T},2,\mu_2}) = \beta_2 > 0. \tag{3.53}$$

From a practical point of view, the two tradeoff filters derived here are fundamentally the same since by taking $\mu = 1/\mu_2$, we see that $\mathbf{H}_{\mathrm{t,T},\mu} = \mathbf{H}_{\mathrm{t,T},2,1/\mu}$.

3.5 SUBSPACE-TYPE FILTER

In [21], it is shown that two symmetric matrices $\mathbf{R_x}$ and $\mathbf{R_v}$ can be jointly diagonalized if $\mathbf{R_v}$ is positive definite. This joint diagonalization was first introduced by Jensen et al. [33] and then by Hu and Loizou [27], [28], [29] in the single-channel noise reduction problem. For our time-domain model, we get

$$\mathbf{R_x} = \mathbf{B}\mathbf{\Lambda}_{\mathrm{jd}}\mathbf{B}^T, \tag{3.54}$$

$$\mathbf{R_v} = \mathbf{B}\mathbf{B}^T, \tag{3.55}$$

$$\mathbf{R_y} = \mathbf{B}\left[\mathbf{I} + \mathbf{\Lambda}_{\mathrm{jd}}\right]\mathbf{B}^T, \tag{3.56}$$

where $\mathbf{B}$ is a full rank square matrix but not necessarily orthogonal, and the diagonal matrix

$$\mathbf{\Lambda}_{\mathrm{jd}} = \mathrm{diag}\left(\lambda_{\mathrm{jd},1}, \lambda_{\mathrm{jd},2}, \ldots, \lambda_{\mathrm{jd},L}\right) \tag{3.57}$$

contains the eigenvalues of the matrix $\mathbf{R_v}^{-1}\mathbf{R_x}$ with $\lambda_{\mathrm{jd},1} \geq \lambda_{\mathrm{jd},2} \geq \cdots \geq \lambda_{\mathrm{jd},L} \geq 0$.

Applying the decompositions (3.54)–(3.56) in (3.47), the tradeoff filter becomes

$$\mathbf{H}_{\mathrm{t,T},\mu} = \mathbf{B}\mathbf{\Lambda}_{\mathrm{jd}}\left(\mathbf{\Lambda}_{\mathrm{jd}} + \mu\mathbf{I}\right)^{-1}\mathbf{B}^{-1}. \tag{3.58}$$

Therefore, the estimation of the speech signal, $\mathbf{x}(m)$, is done in three steps: first, we apply the transform $\mathbf{B}^{-1}$ to the noisy signal; second, the transformed signal is modified by the gain function $\mathbf{\Lambda}_{\mathrm{jd}}\left(\mathbf{\Lambda}_{\mathrm{jd}} + \mu\mathbf{I}\right)^{-1}$; and, finally, we transform back the signal to its original domain by applying the transform $\mathbf{B}$.

It is believed that a speech signal can be modelled as a linear combination of a number of some (linearly independent) basis vectors smaller than the dimension of these vectors [14], [20], [26], [31]. As a result, the vector space of the noisy signal can be decomposed in two subspaces: the signal-plus-noise subspace of length L_{s} and the null subspace of length L_{n}, with $L = L_{\mathrm{s}} + L_{\mathrm{n}}$. This implies that the last L_{n} eigenvalues of the matrix $\mathbf{R_v}^{-1}\mathbf{R_x}$ are equal to zero. Therefore, we can rewrite (3.58) to obtain the subspace-type filter:

$$\mathbf{H}_{\mathrm{t,S},\mu} = \mathbf{B}\begin{bmatrix} \mathbf{\Sigma}_{\mu} & \mathbf{0}_{L_{\mathrm{s}}\times L_{\mathrm{n}}} \\ \mathbf{0}_{L_{\mathrm{n}}\times L_{\mathrm{s}}} & \mathbf{0}_{L_{\mathrm{n}}\times L_{\mathrm{n}}} \end{bmatrix}\mathbf{B}^{-1}, \tag{3.59}$$

where

$$\boldsymbol{\Sigma}_\mu = \text{diag}\left(\frac{\lambda_{\text{jd},1}}{\lambda_{\text{jd},1}+\mu}, \frac{\lambda_{\text{jd},2}}{\lambda_{\text{jd},2}+\mu}, \ldots, \frac{\lambda_{\text{jd},L_\text{s}}}{\lambda_{\text{jd},L_\text{s}}+\mu}\right) \tag{3.60}$$

is an $L_\text{s} \times L_\text{s}$ diagonal matrix. This algorithm is now often referred to as the generalized subspace approach. One should note, however, that there is no noise-only subspace with this formulation. Therefore, noise reduction can only be achieved by modifying the speech-plus-noise subspace by setting μ to a positive number.

Using (3.58) in (3.7), we find that

$$\text{oSNR}(\mathbf{H}_{\text{t},\text{T},\mu}) = \frac{\text{tr}\left[\mathbf{B}\boldsymbol{\Lambda}_{\text{jd}}^3\left(\boldsymbol{\Lambda}_{\text{jd}}+\mu\mathbf{I}\right)^{-2}\mathbf{B}^T\right]}{\text{tr}\left[\mathbf{B}\boldsymbol{\Lambda}_{\text{jd}}^2\left(\boldsymbol{\Lambda}_{\text{jd}}+\mu\mathbf{I}\right)^{-2}\mathbf{B}^T\right]}. \tag{3.61}$$

As a result,

$$\lim_{\mu\to\infty} \text{oSNR}(\mathbf{H}_{\text{t},\text{T},\mu}) = \frac{\text{tr}\left(\mathbf{B}\boldsymbol{\Lambda}_{\text{jd}}^3\mathbf{B}^T\right)}{\text{tr}\left(\mathbf{B}\boldsymbol{\Lambda}_{\text{jd}}^2\mathbf{B}^T\right)}. \tag{3.62}$$

In this limiting case, the tradeoff filter has no interest since $\mathbf{H}_{\text{t},\text{T},\infty} = \mathbf{0}$.

3.6 MAXIMUM SIGNAL-TO-NOISE RATIO (SNR) FILTER

Contrary to what it may be believed, the filtering matrix $\mathbf{B}^{-1}$ that jointly diagonalizes the two matrices $\mathbf{R}_\mathbf{x}$ and $\mathbf{R}_\mathbf{v}$ does not maximize the output SNR.

To derive the maximum SNR filter, we first need to rewrite the filtering matrix as

$$\mathbf{H}_\text{t} = \begin{bmatrix} \mathbf{h}_{\text{t},1}^T \\ \mathbf{h}_{\text{t},2}^T \\ \vdots \\ \mathbf{h}_{\text{t},L}^T \end{bmatrix}, \tag{3.63}$$

where $\mathbf{h}_{\text{t},l}$ is a finite-impulse-response (FIR) filter of length L. We can rewrite the output SNR as

$$\text{oSNR}\left(\mathbf{H}_\text{t}\right) = \frac{\sum_{l=1}^{L} \mathbf{h}_{\text{t},l}^T \mathbf{R}_\mathbf{x} \mathbf{h}_{\text{t},l}}{\sum_{l=1}^{L} \mathbf{h}_{\text{t},l}^T \mathbf{R}_\mathbf{v} \mathbf{h}_{\text{t},l}}. \tag{3.64}$$

Lemma 3.4 *We have*

$$\text{oSNR}\left(\mathbf{H}_\text{t}\right) \le \max_l \frac{\mathbf{h}_{\text{t},l}^T \mathbf{R}_\mathbf{x} \mathbf{h}_{\text{t},l}}{\mathbf{h}_{\text{t},l}^T \mathbf{R}_\mathbf{v} \mathbf{h}_{\text{t},l}} = \chi. \tag{3.65}$$

Proof. Let us define the positive reals $a_l = \mathbf{h}_{\mathrm{t},l}^T \mathbf{R}_{\mathbf{x}} \mathbf{h}_{\mathrm{t},l}$ and $b_l = \mathbf{h}_{\mathrm{t},l}^T \mathbf{R}_{\mathbf{v}} \mathbf{h}_{\mathrm{t},l}$. We have

$$\frac{\sum_{l=1}^{L} a_l}{\sum_{l=1}^{L} b_l} = \sum_{l=1}^{L} \left(\frac{a_l}{b_l} \cdot \frac{b_l}{\sum_{i=1}^{L} b_i} \right). \tag{3.66}$$

Now, define the two following vectors:

$$\mathbf{u} = \begin{bmatrix} \frac{a_1}{b_1} & \frac{a_2}{b_2} & \cdots & \frac{a_L}{b_L} \end{bmatrix}^T, \tag{3.67}$$

$$\mathbf{u}' = \begin{bmatrix} \frac{b_1}{\sum_{i=1}^{L} b_i} & \frac{b_2}{\sum_{i=1}^{L} b_i} & \cdots & \frac{b_L}{\sum_{i=1}^{L} b_i} \end{bmatrix}^T. \tag{3.68}$$

Using the Holder's inequality, we see that

$$\begin{aligned} \frac{\sum_{l=1}^{L} a_l}{\sum_{l=1}^{L} b_l} &= \mathbf{u}^T \mathbf{u}' \\ &\leq \|\mathbf{u}\|_\infty \left\|\mathbf{u}'\right\|_1 = \max_l \frac{a_l}{b_l}, \end{aligned} \tag{3.69}$$

which ends the proof. □

Theorem 3.5 *The maximum SNR filtering matrix is given by*

$$\mathbf{H}_{\mathrm{t,max}} = \begin{bmatrix} \beta_1 \mathbf{h}_{\mathrm{t,max}}^T \\ \beta_2 \mathbf{h}_{\mathrm{t,max}}^T \\ \vdots \\ \beta_L \mathbf{h}_{\mathrm{t,max}}^T \end{bmatrix}, \tag{3.70}$$

where β_l, $l = 1, 2, \ldots, L$ *are real numbers with at least one of them different from* 0 *and* $\mathbf{h}_{\mathrm{t,max}}$ *is the eigenvector corresponding to the maximum eigenvalue,* $\lambda_{\max}$*, of the matrix* $\mathbf{R}_{\mathbf{v}}^{-1}\mathbf{R}_{\mathbf{x}}$. *The corresponding output SNR is*

$$\mathrm{oSNR}\left(\mathbf{H}_{\mathrm{t,max}}\right) = \lambda_{\max}. \tag{3.71}$$

Proof. From Lemma 3.4, we know that the output SNR is upper bounded by χ whose maximum value is clearly $\lambda_{\max}$. On the other hand, it can be checked from (3.64) that $\mathrm{oSNR}\left(\mathbf{H}_{\mathrm{t,max}}\right) = \lambda_{\max}$. Since this output SNR is maximal, $\mathbf{H}_{\mathrm{t,max}}$ is indeed the maximum SNR filter. □

It can be shown that for $\mu \geq 1$,

$$\text{iSNR} \leq \text{oSNR}\left(\mathbf{H}_{\text{t,W}}\right) \leq \text{oSNR}\left(\mathbf{H}_{\text{t,T},\mu}\right) \leq \text{oSNR}\left(\mathbf{H}_{\text{t,max}}\right) = \lambda_{\max} \tag{3.72}$$

and for $\mu \leq 1$,

$$\text{iSNR} \leq \text{oSNR}\left(\mathbf{H}_{\text{t,T},\mu}\right) \leq \text{oSNR}\left(\mathbf{H}_{\text{t,W}}\right) \leq \text{oSNR}\left(\mathbf{H}_{\text{t,max}}\right) = \lambda_{\max}. \tag{3.73}$$

Note that the filtering matrix

$$\mathbf{H}'_{\text{t,max}} = \mathbf{Q}\mathbf{H}_{\text{t,max}} \tag{3.74}$$

also maximizes the output SNR, so that $\mathbf{H}_{\text{t,max}}$ and $\mathbf{H}'_{\text{t,max}}$ are fundamentally equivalent, following the basic principle of maximizing the time-domain output SNR.

CHAPTER 4

Linear Models for Signal Enhancement in the KLE Domain

From the KLE-domain signal model explained in Chapter 2, there are four possible linear models for the estimation of the desired signal as explained in this part.

4.1 MODEL 1

In the first and simplest model, that we call Model 1, neither interframe nor interband correlations are taken into account. With this model, the estimate of $c_{x,l}(m)$ is obtained with

$$\begin{aligned} c_{z_1,l}(m) &= h_{1,l}c_{y,l}(m) \\ &= h_{1,l}c_{x,l}(m) + h_{1,l}c_{v,l}(m),\ l = 1, 2, \ldots, L, \end{aligned} \tag{4.1}$$

where $h_{1,l}$ is a (positive) gain factor that should be smaller than 1. This approach is pretty much equivalent to noise reduction in the frequency domain [7], which ignores the interband and interframe correlations of the signals.

The variance of $c_{z_1,l}(m)$ is

$$\begin{aligned} \phi_{c_{z_1,l}} &= E\left[c^2_{z_1,l}(m)\right] \\ &= h^2_{1,l}\phi_{c_{y,l}} \\ &= h^2_{1,l}\lambda_l \\ &= h^2_{1,l}\phi_{c_{x,l}} + h^2_{1,l}\phi_{c_{v,l}},\ l = 1, 2, \ldots, L, \end{aligned} \tag{4.2}$$

where

$$\phi_{c_{y,l}} = \lambda_l, \tag{4.3}$$

$$\phi_{c_{x,l}} = \mathbf{q}_l^T \mathbf{R}_\mathbf{x} \mathbf{q}_l, \tag{4.4}$$

$$\phi_{c_{v,l}} = \mathbf{q}_l^T \mathbf{R}_\mathbf{v} \mathbf{q}_l, \tag{4.5}$$

are the variances of $c_{y,l}(m)$, $c_{x,l}(m)$, and $c_{v,l}(m)$, respectively. Intuitively, we see from (4.2) that for the eigenvalues dominated by noise, the corresponding gains should be close to 0, while for the eigenvalues dominated by speech, the corresponding gains should be close to 1.

With Model 1, we can deduce the estimate of $\mathbf{x}(m)$ as

$$\begin{aligned}
\mathbf{z}_1(m) &= \sum_{l=1}^{L} c_{z_1,l}(m)\mathbf{q}_l \\
&= \left(\sum_{l=1}^{L} h_{1,l}\mathbf{q}_l\mathbf{q}_l^T\right)\mathbf{y}(m) \\
&= \mathbf{H}_{\mathrm{TD},1}\mathbf{y}(m),
\end{aligned} \tag{4.6}$$

where

$$\begin{aligned}
\mathbf{H}_{\mathrm{TD},1} &= \sum_{l=1}^{L} h_{1,l}\mathbf{q}_l\mathbf{q}_l^T \\
&= \mathbf{Q}\,\mathrm{diag}\left(h_{1,1}, h_{1,2}, \ldots, h_{1,L}\right)\mathbf{Q}^T
\end{aligned} \tag{4.7}$$

is a matrix of size $L \times L$, which is the equivalent time-domain version of the gains $h_{1,l}$ in the KLE domain. Hence, the correlation matrix of $\mathbf{z}_1(m)$ is

$$\mathbf{R}_{\mathbf{z}_1} = \mathbf{Q}\,\mathrm{diag}\left(h_{1,1}^2\lambda_1, h_{1,2}^2\lambda_2, \ldots, h_{1,L}^2\lambda_L\right)\mathbf{Q}^T. \tag{4.8}$$

4.2 MODEL 2

In Model 2, the interframe correlation is taken into account. Therefore, we estimate the coefficients $c_{x,l}(m),\ l = 1, 2, \ldots, L$, by passing $c_{y,l}(m),\ l = 1, 2, \ldots, L$, from consecutive time-frames through a linear filter, i.e.,

$$\begin{aligned}
c_{z_2,l}(m) &= \mathbf{h}_{2,l}^T\mathbf{c}_{y,l}(m) \\
&= \mathbf{h}_{2,l}^T\mathbf{c}_{x,l}(m) + \mathbf{h}_{2,l}^T\mathbf{c}_{v,l}(m), \quad l = 1, 2, \ldots, L,
\end{aligned} \tag{4.9}$$

where

$$\mathbf{h}_{2,l} = \begin{bmatrix} h_{2,l,0} & h_{2,l,1} & \cdots & h_{2,l,M-1} \end{bmatrix}^T$$

is an FIR filter of length M corresponding to the subband l,

$$\mathbf{c}_{y,l}(m) = \begin{bmatrix} c_{y,l}(m) & c_{y,l}(m-1) & \cdots & c_{y,l}(m-M+1) \end{bmatrix}^T$$

is a vector of length M, $\mathbf{c}_{x,l}(m)$ and $\mathbf{c}_{v,l}(m)$ are defined in a similar way to $\mathbf{c}_{y,l}(m)$, and M is the chosen number of consecutive frames. Taking $M = 1$, for all the filters $\mathbf{h}_{2,l}$ in (4.9), we get Model 1 presented in the previous subsection. However, for $M > 1$, the interframe correlation will now be taken into account.

At time-frame m, our desired signal is $c_{x,l}(m)$ [and not the whole the vector $\mathbf{c}_{x,l}(m)$]. However, the vector $\mathbf{c}_{x,l}(m)$ contains both the desired signal, $c_{x,l}(m)$, and the components $c_{x,l}(m-i),\ i \neq 0,$

which are not the desired signals at time-frame m but signals that are correlated with $c_{x,l}(m)$. Therefore, the elements $c_{x,l}(m-i),\ i \neq 0$, contain both a part of the desired signal and a component that we consider as an interference. This suggests that we should decompose $c_{x,l}(m-i)$ into two orthogonal components corresponding to the part of the desired signal and interference, i.e.,

$$c_{x,l}(m-i) = \gamma_{c_{x,l}}(i) c_{x,l}(m) + c'_{x,l}(m-i), \tag{4.10}$$

where

$$c'_{x,l}(m-i) = c_{x,l}(m-i) - \gamma_{c_{x,l}}(i) c_{x,l}(m), \tag{4.11}$$

$$E\left[c_{x,l}(m) c'_{x,l}(m-i)\right] = 0, \tag{4.12}$$

and

$$\gamma_{c_{x,l}}(i) = \frac{E\left[c_{x,l}(m) c_{x,l}(m-i)\right]}{E\left[c^2_{x,l}(m)\right]} \tag{4.13}$$

is the interframe correlation coefficient of the signal $c_{x,l}(m)$. Hence, we can write the vector $\mathbf{c}_{x,l}(m)$ as

$$\begin{aligned} \mathbf{c}_{x,l}(m) &= c_{x,l}(m) \boldsymbol{\gamma}_{c_{x,l}} + \mathbf{c}'_{x,l}(m) \\ &= \mathbf{c}_{x_{\mathrm{d}},l}(m) + \mathbf{c}'_{x,l}(m), \end{aligned} \tag{4.14}$$

where $\mathbf{c}_{x_{\mathrm{d}},l}(m) = c_{x,l}(m) \boldsymbol{\gamma}_{c_{x,l}}$ is a vector depending on the desired signal,

$$\mathbf{c}'_{x,l}(m) = \begin{bmatrix} c'_{x,l}(m) & c'_{x,l}(m-1) & \cdots & c'_{x,l}(m-M+1) \end{bmatrix}^T$$

is the interference signal vector, and

$$\begin{aligned} \boldsymbol{\gamma}_{c_{x,l}} &= \begin{bmatrix} \gamma_{c_{x,l}}(0) & \gamma_{c_{x,l}}(1) & \cdots & \gamma_{c_{x,l}}(M-1) \end{bmatrix}^T \\ &= \begin{bmatrix} 1 & \gamma_{c_{x,l}}(1) & \cdots & \gamma_{c_{x,l}}(M-1) \end{bmatrix}^T \\ &= \frac{E\left[c_{x,l}(m) \mathbf{c}_{x,l}(m)\right]}{E\left[c^2_{x,l}(m)\right]} \end{aligned} \tag{4.15}$$

is the (normalized) interframe correlation vector.

Substituting (4.14) into (4.9), we get

$$c_{z_2,l}(m) = c_{x,l}(m) \mathbf{h}^T_{2,l} \boldsymbol{\gamma}_{c_{x,l}} + \mathbf{h}^T_{2,l} \mathbf{c}'_{x,l}(m) + \mathbf{h}^T_{2,l} \mathbf{c}_{v,l}(m), \quad l = 1, 2, \ldots, L. \tag{4.16}$$

We observe that the estimate of the desired signal is the sum of three terms that are mutually uncorrelated. The first one is clearly the filtered desired signal while the two others are the filtered undesired signals (interference-plus-noise). Therefore, the variance of $c_{z_2,l}(m)$ is

$$\begin{aligned}\phi_{c_{z_2,l}} &= \mathbf{h}_{2,l}^T \mathbf{\Phi}_{\mathbf{c}_{y,l}} \mathbf{h}_{2,l} \\ &= \mathbf{h}_{2,l}^T \mathbf{\Phi}_{\mathbf{c}_{x_d,l}} \mathbf{h}_{2,l} + \mathbf{h}_{2,l}^T \mathbf{\Phi}_{\mathbf{c}'_{x,l}} \mathbf{h}_{2,l} + \mathbf{h}_{2,l}^T \mathbf{\Phi}_{\mathbf{c}_{v,l}} \mathbf{h}_{2,l},\ l = 1, 2, \ldots, L, \end{aligned} \tag{4.17}$$

where

$$\mathbf{\Phi}_{\mathbf{c}_{y,l}} = E\left[\mathbf{c}_{y,l}(m)\mathbf{c}_{y,l}^T(m)\right], \tag{4.18}$$

$$\begin{aligned}\mathbf{\Phi}_{\mathbf{c}_{x_d,l}} &= E\left[\mathbf{c}_{x_d,l}(m)\mathbf{c}_{x_d,l}^T(m)\right] \\ &= \phi_{c_{x,l}} \boldsymbol{\gamma}_{c_{x,l}} \boldsymbol{\gamma}_{c_{x,l}}^T, \end{aligned} \tag{4.19}$$

$$\begin{aligned}\mathbf{\Phi}_{\mathbf{c}'_{x,l}} &= E\left[\mathbf{c}'_{x,l}(m)\mathbf{c}'^T_{x,l}(m)\right] \\ &= \mathbf{\Phi}_{\mathbf{c}_{x,l}} - \mathbf{\Phi}_{\mathbf{c}_{x_d,l}}, \end{aligned} \tag{4.20}$$

$$\mathbf{\Phi}_{\mathbf{c}_{v,l}} = E\left[\mathbf{c}_{v,l}(m)\mathbf{c}_{v,l}^T(m)\right], \tag{4.21}$$

are the correlation matrices of the vectors $\mathbf{c}_{y,l}(m)$, $\mathbf{c}_{x_d,l}(m)$, $\mathbf{c}'_{x,l}(m)$, and $\mathbf{c}_{v,l}(m)$, respectively. We see clearly from these correlation matrices that the interframe correlation is taken into account.

The estimate of the vector $\mathbf{x}(m)$ would be

$$\begin{aligned}\mathbf{z}_2(m) &= \sum_{l=1}^{L} c_{z_2,l}(m)\mathbf{q}_l \\ &= \sum_{l=1}^{L} \sum_{i=0}^{M-1} h_{2,l,i} c_{y,l}(m-i)\mathbf{q}_l \\ &= \sum_{i=0}^{M-1} \sum_{l=1}^{L} h_{2,l,i} \mathbf{q}_l \mathbf{q}_l^T \mathbf{y}(m-i) \\ &= \sum_{i=0}^{M-1} \mathbf{H}_{\mathrm{TD},2,i} \mathbf{y}(m-i), \end{aligned} \tag{4.22}$$

where

$$\mathbf{H}_{\mathrm{TD},2,i} = \sum_{l=1}^{L} h_{2,l,i} \mathbf{q}_l \mathbf{q}_l^T,\ i = 0, 1, \ldots, M-1 \tag{4.23}$$

are the time-domain filtering matrices. We see again from (4.22) how the estimate depends on the M successive frames of the observation signal vector $\mathbf{y}(m)$. The correlation matrix of $\mathbf{z}_2(m)$ is

$$\mathbf{R}_{\mathbf{z}_2} = \sum_{i=0}^{M-1} \sum_{j=0}^{M-1} \mathbf{H}_{\mathrm{TD},2,i} E\left[\mathbf{y}(m-i)\mathbf{y}^T(m-j)\right] \mathbf{H}_{\mathrm{TD},2,j}^T. \tag{4.24}$$

4.3 MODEL 3

In our third model, the interband correlation is taken into account. Then, we have

$$\begin{aligned} c_{z3,l}(m) &= \mathbf{h}_{3,l}^T \mathbf{c}_y(m) \\ &= \mathbf{h}_{3,l}^T \mathbf{c}_x(m) + \mathbf{h}_{3,l}^T \mathbf{c}_v(m), \quad l = 1, 2, \ldots, L, \end{aligned} \tag{4.25}$$

where

$$\mathbf{h}_{3,l} = \begin{bmatrix} h_{3,l,0} & h_{3,l,1} & \cdots & h_{3,l,L'-1} \end{bmatrix}^T$$

is an FIR filter of length $L' \leq L$, corresponding to the subband l,

$$\mathbf{c}_y(m) = \begin{bmatrix} c_{y,1}(m) & c_{y,2}(m) & \cdots & c_{y,L'}(m) \end{bmatrix}^T \tag{4.26}$$

is a vector of length L', and $\mathbf{c}_x(m)$ and $\mathbf{c}_v(m)$ are defined in a similar way to $\mathbf{c}_y(m)$. Taking $L' = 1$ for all the filters $\mathbf{h}_{3,l}$ in (4.25), we obtain Model 1. However, for $L' > 1$, the interband correlation will now be taken into account. In the rest, we will always assume that $L' = L$. In this case, $\mathbf{c}_y(m) = \mathbf{Q}^T\mathbf{y}(m)$, $\mathbf{c}_x(m) = \mathbf{Q}^T\mathbf{x}(m)$, and $\mathbf{c}_v(m) = \mathbf{Q}^T\mathbf{v}(m)$.

In a vector form, (4.25) is

$$\begin{aligned} \mathbf{c}_{z_3}(m) &= \begin{bmatrix} c_{z3,1}(m) & c_{z3,2}(m) & \cdots & c_{z3,L}(m) \end{bmatrix}^T \\ &= \mathbf{H}_3 \mathbf{c}_y(m) \\ &= \mathbf{H}_3 \mathbf{c}_x(m) + \mathbf{H}_3 \mathbf{c}_v(m), \end{aligned} \tag{4.27}$$

where

$$\mathbf{H}_3 = \begin{bmatrix} \mathbf{h}_{3,1}^T \\ \mathbf{h}_{3,2}^T \\ \vdots \\ \mathbf{h}_{3,L}^T \end{bmatrix}$$

is a filtering matrix of size $L \times L$. For this model, $\mathbf{c}_x(m)$ is our desired signal vector.

The correlation matrix of $\mathbf{c}_{z_3}(m)$ is

$$\begin{aligned} \mathbf{\Phi}_{\mathbf{c}_{z_3}} &= \mathbf{H}_3 \mathbf{\Phi}_{\mathbf{c}_y} \mathbf{H}_3^T \\ &= \mathbf{H}_3 \mathbf{\Phi}_{\mathbf{c}_x} \mathbf{H}_3^T + \mathbf{H}_3 \mathbf{\Phi}_{\mathbf{c}_v} \mathbf{H}_3^T, \end{aligned} \tag{4.28}$$

where

$$\begin{aligned} \mathbf{\Phi}_{\mathbf{c}_y} &= E\left[\mathbf{c}_y(m)\mathbf{c}_y^T(m)\right] \\ &= \mathbf{\Lambda}, \end{aligned} \tag{4.29}$$

$$\begin{aligned} \mathbf{\Phi}_{\mathbf{c}_x} &= E\left[\mathbf{c}_x(m)\mathbf{c}_x^T(m)\right] \\ &= \mathbf{Q}^T\mathbf{R}_\mathbf{x}\mathbf{Q}, \end{aligned} \tag{4.30}$$

$$\begin{aligned} \mathbf{\Phi}_{\mathbf{c}_v} &= E\left[\mathbf{c}_v(m)\mathbf{c}_v^T(m)\right] \\ &= \mathbf{Q}^T\mathbf{R}_\mathbf{v}\mathbf{Q}. \end{aligned} \tag{4.31}$$

are the correlation matrices of the vectors $\mathbf{c}_y(m)$, $\mathbf{c}_x(m)$, and $\mathbf{c}_v(m)$, respectively.

With Model 3, the estimate of $\mathbf{x}(m)$ is

$$\begin{aligned}\mathbf{z}_3(m) &= \mathbf{Q}\mathbf{c}_{z_3}(m) \\ &= \mathbf{Q}\mathbf{H}_3\mathbf{Q}^T\mathbf{y}(m) \\ &= \mathbf{H}_{\mathrm{TD},3}\mathbf{y}(m),\end{aligned} \tag{4.32}$$

where

$$\mathbf{H}_{\mathrm{TD},3} = \mathbf{Q}\mathbf{H}_3\mathbf{Q}^T \tag{4.33}$$

is the time-domain form of $\mathbf{H}_3$. Therefore, the correlation matrix of $\mathbf{z}_3(m)$ is

$$\mathbf{R}_{\mathbf{z}_3} = \mathbf{Q}\mathbf{H}_3\mathbf{\Lambda}\mathbf{H}_3^T\mathbf{Q}^T, \tag{4.34}$$

which is interesting to compare to $\mathbf{R}_{\mathbf{z}_1}$ of Model 1.

4.4 MODEL 4

In our fourth and last model, we take into account both the interframe and interband correlations. In this case, the coefficients $c_{x,l}(m),\ l = 1, 2, \ldots, L$, are estimated as

$$\begin{aligned}c_{z4,l}(m) &= \sum_{i=0}^{M-1} \mathbf{h}_{4,l,i}^T\mathbf{c}_y(m-i) \\ &= \sum_{i=0}^{M-1} \mathbf{h}_{4,l,i}^T\mathbf{c}_x(m-i) + \sum_{i=0}^{M-1} \mathbf{h}_{4,l,i}^T\mathbf{c}_v(m-i),\ \ l = 1, 2, \ldots, L,\end{aligned} \tag{4.35}$$

where

$$\mathbf{h}_{4,l,i} = \begin{bmatrix} h_{4,l,i,0} & h_{4,l,i,1} & \cdots & h_{4,l,i,L'-1} \end{bmatrix}^T$$

is an FIR filter of length $L' \leq L$, corresponding to the subband index l and time-frame index i,

$$\mathbf{c}_y(m-i) = \begin{bmatrix} c_{y,1}(m-i) & c_{y,2}(m-i) & \cdots & c_{y,L'}(m-i) \end{bmatrix}^T \tag{4.36}$$

is a vector of length L', and $\mathbf{c}_x(m-i)$ and $\mathbf{c}_v(m-i)$ are defined in a similar way to $\mathbf{c}_y(m-i)$. Model 4 is a generalization of the three previous models. Indeed, taking $L' = M = 1$ for all, the filters in (4.35) gives Model 1; $L' = 1$ leads to Model 2; and $M = 1$ corresponds to Model 3. In the rest, we will always assume that $L' = L$.

Expression (4.35) can be rewritten in a more convenient way as

$$\begin{aligned}c_{z4,l}(m) &= \underline{\mathbf{h}}_{4,l}^T\underline{\mathbf{c}}_y(m) \\ &= \underline{\mathbf{h}}_{4,l}^T\underline{\mathbf{c}}_x(m) + \underline{\mathbf{h}}_{4,l}^T\underline{\mathbf{c}}_v(m),\ \ l = 1, 2, \ldots, L,\end{aligned} \tag{4.37}$$

where

$$\underline{\mathbf{h}}_{4,l} = \begin{bmatrix} \mathbf{h}_{4,l,0}^T & \mathbf{h}_{4,l,1}^T & \cdots & \mathbf{h}_{4,l,M-1}^T \end{bmatrix}^T$$

is an FIR filter of length ML,

$$\underline{\mathbf{c}}_y(m) = \begin{bmatrix} \mathbf{c}_y^T(m) & \mathbf{c}_y^T(m-1) & \cdots & \mathbf{c}_y^T(m-M+1) \end{bmatrix}^T$$

is a vector of length ML, and $\underline{\mathbf{c}}_x(m)$ and $\underline{\mathbf{c}}_v(m)$ are defined in a similar way to $\underline{\mathbf{c}}_y(m)$.

In a vector form, (4.37) becomes

$$\begin{aligned} \mathbf{c}_{z4}(m) &= \begin{bmatrix} c_{z4,1}(m) & c_{z4,2}(m) & \cdots & c_{z4,L}(m) \end{bmatrix}^T \\ &= \underline{\mathbf{H}}_4 \underline{\mathbf{c}}_y(m) \\ &= \underline{\mathbf{H}}_4 \underline{\mathbf{c}}_x(m) + \underline{\mathbf{H}}_4 \underline{\mathbf{c}}_v(m), \end{aligned} \tag{4.38}$$

where

$$\underline{\mathbf{H}}_4 = \begin{bmatrix} \underline{\mathbf{h}}_{4,1}^T \\ \underline{\mathbf{h}}_{4,2}^T \\ \vdots \\ \underline{\mathbf{h}}_{4,L}^T \end{bmatrix}$$

is a filtering matrix of size $L \times ML$.

At time-frame m, our desired signal vector is $\mathbf{c}_x(m)$ but not the whole vector $\underline{\mathbf{c}}_x(m)$. Therefore, we should decompose $\underline{\mathbf{c}}_x(m)$ into two orthogonal components:

$$\begin{aligned} \underline{\mathbf{c}}_x(m) &= \underline{\mathbf{\Gamma}}_{\mathbf{c}_x} \mathbf{c}_x(m) + \underline{\mathbf{c}}_x''(m) \\ &= \underline{\mathbf{c}}_{x_\mathrm{d}}(m) + \underline{\mathbf{c}}_x''(m), \end{aligned} \tag{4.39}$$

where $\underline{\mathbf{c}}_{x_\mathrm{d}}(m) = \underline{\mathbf{\Gamma}}_{\mathbf{c}_x} \mathbf{c}_x(m)$ is a linear version of the desired signal vector, $\underline{\mathbf{c}}_x''(m)$ is the interference signal vector of length ML,

$$\underline{\mathbf{\Gamma}}_{\mathbf{c}_x} = \begin{bmatrix} \mathbf{\Phi}_{\mathbf{c}_x,0} \mathbf{\Phi}_{\mathbf{c}_x}^{-1} \\ \mathbf{\Phi}_{\mathbf{c}_x,1} \mathbf{\Phi}_{\mathbf{c}_x}^{-1} \\ \vdots \\ \mathbf{\Phi}_{\mathbf{c}_x,M-1} \mathbf{\Phi}_{\mathbf{c}_x}^{-1} \end{bmatrix}$$

is the normalized interframe correlation matrix,

$$\mathbf{\Phi}_{\mathbf{c}_x,i} = E\left[\mathbf{c}_x(m-i)\mathbf{c}_x^T(m)\right], \ i = 0, 1, \ldots, M-1, \tag{4.40}$$

and

$$E\left[\mathbf{c}_x(m)\underline{\mathbf{c}}_x''^T(m)\right] = \mathbf{0}. \tag{4.41}$$

Substituting (4.39) in (4.38), we obtain

$$\mathbf{c}_{z_4}(m) = \underline{\mathbf{H}}_4 \underline{\boldsymbol{\Gamma}}_{\mathbf{c}_x} \mathbf{c}_x(m) + \underline{\mathbf{H}}_4 \underline{\mathbf{c}}_x''(m) + \underline{\mathbf{H}}_4 \underline{\mathbf{c}}_v(m) \tag{4.42}$$

and the correlation matrix of $\mathbf{c}_{z_4}(m)$ is

$$\boldsymbol{\Phi}_{\mathbf{c}_{z_4}} = \underline{\mathbf{H}}_4 \boldsymbol{\Phi}_{\underline{\mathbf{c}}_{x_\mathrm{d}}} \underline{\mathbf{H}}_4^T + \underline{\mathbf{H}}_4 \boldsymbol{\Phi}_{\underline{\mathbf{c}}_x''} \underline{\mathbf{H}}_4^T + \underline{\mathbf{H}}_4 \boldsymbol{\Phi}_{\underline{\mathbf{c}}_v} \underline{\mathbf{H}}_4^T, \tag{4.43}$$

where

$$\begin{aligned} \boldsymbol{\Phi}_{\underline{\mathbf{c}}_{x_\mathrm{d}}} &= E\left[\underline{\mathbf{c}}_{x_\mathrm{d}}(m)\underline{\mathbf{c}}_{x_\mathrm{d}}^T(m)\right] \\ &= \underline{\boldsymbol{\Gamma}}_{\mathbf{c}_x} \boldsymbol{\Phi}_{\mathbf{c}_x} \underline{\boldsymbol{\Gamma}}_{\mathbf{c}_x}^T, \end{aligned} \tag{4.44}$$

$$\begin{aligned} \boldsymbol{\Phi}_{\underline{\mathbf{c}}_x''} &= E\left[\underline{\mathbf{c}}_x''(m)\underline{\mathbf{c}}_x''^T(m)\right] \\ &= \boldsymbol{\Phi}_{\underline{\mathbf{c}}_x} - \boldsymbol{\Phi}_{\underline{\mathbf{c}}_{x_\mathrm{d}}}, \end{aligned} \tag{4.45}$$

$$\boldsymbol{\Phi}_{\underline{\mathbf{c}}_v} = E\left[\underline{\mathbf{c}}_v(m)\underline{\mathbf{c}}_v^T(m)\right], \tag{4.46}$$

are the correlation matrices of the vectors $\underline{\mathbf{c}}_{x_\mathrm{d}}(m)$, $\underline{\mathbf{c}}_x''(m)$, and $\underline{\mathbf{c}}_v(m)$, respectively.

With this model, the estimate of $\mathbf{x}(m)$ is

$$\begin{aligned} \mathbf{z}_4(m) &= \sum_{l=1}^{L} c_{z_4,l}(m)\mathbf{q}_l \\ &= \sum_{l=1}^{L}\sum_{i=0}^{M-1} \mathbf{h}_{4,l,i}^T \mathbf{c}_y(m-i)\mathbf{q}_l \\ &= \sum_{i=0}^{M-1}\sum_{l=1}^{L} \mathbf{q}_l \mathbf{h}_{4,l,i}^T \mathbf{Q}^T \mathbf{y}(m-i) \\ &= \sum_{i=0}^{M-1} \mathbf{H}_{\mathrm{TD},4,i}\mathbf{y}(m-i), \end{aligned} \tag{4.47}$$

where

$$\mathbf{H}_{\mathrm{TD},4,i} = \sum_{l=1}^{L} \mathbf{q}_l \mathbf{h}_{4,l,i}^T \mathbf{Q}^T, \; i = 0, 1, \ldots, M-1 \tag{4.48}$$

are the time-domain filtering matrices. The correlation matrix of $\mathbf{z}_4(m)$ is

$$\mathbf{R}_{\mathbf{z}_4} = \sum_{i=0}^{M-1}\sum_{j=0}^{M-1} \mathbf{H}_{\mathrm{TD},4,i} E\left[\mathbf{y}(m-i)\mathbf{y}^T(m-j)\right] \mathbf{H}_{\mathrm{TD},4,j}^T. \tag{4.49}$$

CHAPTER 5

Optimal Filters in the KLE Domain with Model 1

In this chapter, we study noise reduction with Model 1. We recall that in Model 1, neither interframe nor interband correlations are taken into account. To simplify the presentation, we drop the subscript "1" from the gain (see Chapter 4, Section 4.1), so that now $h_{1,l}$ is written as h_l.

5.1 PERFORMANCE MEASURES

To examine what happens in each subband, we define the subband input SNR as

$$\begin{aligned} \text{iSNR}_l &= \frac{\phi_{c_{x,l}}}{\phi_{c_{v,l}}} \\ &= \frac{\mathbf{q}_l^T \mathbf{R}_\mathbf{x} \mathbf{q}_l}{\mathbf{q}_l^T \mathbf{R}_\mathbf{v} \mathbf{q}_l},\ l = 1, 2, \ldots, L. \end{aligned} \tag{5.1}$$

We can rewrite the input SNR (already defined in Chapter 3) as

$$\begin{aligned} \text{iSNR} &= \frac{\sum_{l=1}^{L} \mathbf{q}_l^T \mathbf{R}_\mathbf{x} \mathbf{q}_l}{\sum_{l=1}^{L} \mathbf{q}_l^T \mathbf{R}_\mathbf{v} \mathbf{q}_l} \\ &= \frac{\sigma_x^2}{\sigma_v^2}. \end{aligned} \tag{5.2}$$

We can demonstrate that [7]

$$\text{iSNR} \le \sum_{l=1}^{L} \text{iSNR}_l. \tag{5.3}$$

The output SNR is the SNR after the filtering operation. From (4.2), we deduce the subband output SNR:

$$\begin{aligned} \text{oSNR}(h_l) &= \frac{h_l^2 \phi_{c_{x,l}}}{h_l^2 \phi_{c_{v,l}}} \\ &= \text{iSNR}_l,\ l = 1, 2, \ldots, L \end{aligned} \tag{5.4}$$

and the fullband output SNR:

$$\text{oSNR}(h_:) = \frac{\sum_{l=1}^{L} h_l^2 \phi_{c_{x,l}}}{\sum_{l=1}^{L} h_l^2 \phi_{c_{v,l}}}. \tag{5.5}$$

We notice that the subband output SNR cannot be improved with just a gain, but the fullband output SNR can. We always have [7]

$$\text{oSNR}(h_:) \leq \sum_{l=1}^{L} \text{iSNR}_l. \tag{5.6}$$

The previous inequality shows that the fullband output SNR is upper bounded no matter how the gains $h_l,\ l = 1, 2, \ldots, L$ are chosen.

The subband and fullband noise-reduction factors are

$$\begin{aligned} \xi_{\text{nr}}(h_l) &= \frac{\phi_{c_{v,l}}}{h_l^2 \phi_{c_{v,l}}} \\ &= \frac{1}{h_l^2},\ l = 1, 2, \ldots, L, \end{aligned} \tag{5.7}$$

$$\begin{aligned} \xi_{\text{nr}}(h_:) &= \frac{\sum_{l=1}^{L} \phi_{c_{v,l}}}{\sum_{l=1}^{L} h_l^2 \phi_{c_{v,l}}} \\ &= \frac{\sum_{l=1}^{L} \phi_{c_{v,l}}}{\sum_{l=1}^{L} \xi_{\text{nr}}^{-1}(h_l) \phi_{c_{v,l}}}. \end{aligned} \tag{5.8}$$

The noise-reduction factor is supposed to have a lower bound of 1 for optimal gains, and the larger its value, the more the noise is reduced. We also have

$$\xi_{\text{nr}}(h_:) \leq \sum_{l=1}^{L} \xi_{\text{nr}}(h_l). \tag{5.9}$$

To quantify the speech distortion, we give the subband speech-distortion index

$$\begin{aligned} \upsilon_{\text{sd}}(h_l) &= \frac{E\left\{\left[h_l c_{x,l}(m) - c_{x,l}(m)\right]^2\right\}}{\phi_{c_{x,l}}} \\ &= (h_l - 1)^2,\ l = 1, 2, \ldots, L \end{aligned} \tag{5.10}$$

and the fullband speech-distortion index

$$\begin{aligned} \upsilon_{\text{sd}}(h_:) &= \frac{\sum_{l=1}^{L} E\left\{\left[h_l c_{x,l}(m) - c_{x,l}(m)\right]^2\right\}}{\sum_{l=1}^{L} \phi_{c_{x,l}}} \\ &= \frac{\sum_{l=1}^{L} \upsilon_{\text{sd}}(h_l) \phi_{c_{x,l}}}{\sum_{l=1}^{L} \phi_{c_{x,l}}}. \end{aligned} \tag{5.11}$$

The speech-distortion index is usually upper bounded by 1. We have

$$\upsilon_{sd}(h_:) \leq \sum_{l=1}^{L} \upsilon_{sd}(h_l). \tag{5.12}$$

Another way to quantify signal distortion is via the speech-reduction factor. The subband and fullband definitions are

$$\begin{aligned} \xi_{sr}(h_l) &= \frac{\phi_{c_{x,l}}}{h_l^2 \phi_{c_{x,l}}} \\ &= \frac{1}{h_l^2},\ l = 1, 2, \ldots, L, \end{aligned} \tag{5.13}$$

$$\begin{aligned} \xi_{sr}(h_:) &= \frac{\sum_{l=1}^{L} \phi_{c_{x,l}}}{\sum_{l=1}^{L} h_l^2 \phi_{c_{x,l}}} \\ &= \frac{\sum_{l=1}^{L} \phi_{c_{x,l}}}{\sum_{l=1}^{L} \xi_{sr}^{-1}(h_l) \phi_{c_{x,l}}}. \end{aligned} \tag{5.14}$$

The speech-reduction factor is supposed to have a lower bound of 1 for optimal gains. We also have

$$\xi_{sr}(h_:) \leq \sum_{l=1}^{L} \xi_{sr}(h_l). \tag{5.15}$$

It can easily be checked that

$$\frac{\text{oSNR}(h_l)}{\text{iSNR}_l} = \frac{\xi_{nr}(h_l)}{\xi_{sr}(h_l)},\ l = 1, 2, \ldots, L, \tag{5.16}$$

$$\frac{\text{oSNR}(h_:)}{\text{iSNR}} = \frac{\xi_{nr}(h_:)}{\xi_{sr}(h_:)}. \tag{5.17}$$

5.2 MSE CRITERION

In the KLE domain and with Model 1, the error signal between the estimated and desired signals in the subband l is

$$\begin{aligned} e_l(m) &= c_{z_1,l}(m) - c_{x,l}(m) \\ &= h_l c_{y,l}(m) - c_{x,l}(m), \end{aligned} \tag{5.18}$$

which can also be written as the sum of two uncorrelated error signals:

$$e_l(m) = e_{x,l}(m) + e_{v,l}(m), \tag{5.19}$$

where

$$e_{x,l}(m) = h_l c_{x,l}(m) - c_{x,l}(m) \tag{5.20}$$

is the speech distortion due to the gain and

$$e_{v,l}(m) = h_l c_{v,l}(m) \tag{5.21}$$

represents the residual noise.

From the error signal (5.18), we give the corresponding KLE-domain (or subband) MSE criterion:

$$\begin{aligned} J(h_l) &= E\left[e_l^2(m)\right] \\ &= h_l^2 \lambda_l - 2h_l \phi_{c_{x,l}c_{y,l}} + \phi_{c_{x,l}}, \end{aligned} \tag{5.22}$$

where

$$\begin{aligned} \phi_{c_{x,l}c_{y,l}} &= E\left[c_{x,l}(m)c_{y,l}(m)\right] \\ &= E\left[c_{x,l}^2(m)\right] \\ &= \phi_{c_{x,l}} \end{aligned}$$

is the cross-correlation between the signals $c_{x,l}(m)$ and $c_{y,l}(m)$. Expression (5.22) can be structured in a different way:

$$\begin{aligned} J(h_l) &= E\left[e_{x,l}^2(m)\right] + E\left[e_{v,l}^2(m)\right] \\ &= J_x(h_l) + J_v(h_l). \end{aligned} \tag{5.23}$$

For the particular gain $h_l = 1, \ \forall l$, we get

$$\begin{aligned} J(1) &= E\left[c_{v,l}^2(m)\right] \\ &= \phi_{c_{v,l}} \\ &= \mathbf{q}_l^T \mathbf{R_v} \mathbf{q}_l, \end{aligned} \tag{5.24}$$

so there will be neither noise reduction nor speech distortion. Using this particular case of the MSE, we define the KLE-domain (or subband) normalized MSE (NMSE) as

$$\begin{aligned} \tilde{J}(h_l) &= \frac{J(h_l)}{J(1)} \\ &= \text{iSNR}_l \cdot \upsilon_{\text{sd}}(h_l) + \frac{1}{\xi_{\text{nr}}(h_l)}, \end{aligned} \tag{5.25}$$

where

$$\upsilon_{\text{sd}}(h_l) = \frac{J_x(h_l)}{\phi_{c_{x,l}}}, \tag{5.26}$$

$$\xi_{\text{nr}}(h_l) = \frac{\mathbf{q}_l^T \mathbf{R}_\mathbf{v} \mathbf{q}_l}{J_v(h_l)}. \tag{5.27}$$

The KLE-domain NMSE depends explicitly on the subband speech-distortion index and the subband noise-reduction factor.

We define the fullband MSE and fullband NMSE as

$$\begin{aligned} J(h_:) &= \frac{1}{L}\sum_{l=1}^{L} J(h_l) \\ &= \frac{1}{L}\sum_{l=1}^{L}(h_l - 1)^2 \phi_{c_{x,l}} + \frac{1}{L}\sum_{l=1}^{L} h_l^2 \phi_{c_{v,l}} \\ &= J_x(h_:) + J_v(h_:) \end{aligned} \tag{5.28}$$

and

$$\begin{aligned} \tilde{J}(h_:) &= L\frac{J(h_:)}{\sum_{l=1}^{L} \mathbf{q}_l^T \mathbf{R}_\mathbf{v} \mathbf{q}_l} \\ &= \frac{\sum_{l=1}^{L}(h_l - 1)^2 \phi_{c_{x,l}}}{\sum_{l=1}^{L} \mathbf{q}_l^T \mathbf{R}_\mathbf{v} \mathbf{q}_l} + \frac{\sum_{l=1}^{L} h_l^2 \phi_{c_{v,l}}}{\sum_{l=1}^{L} \mathbf{q}_l^T \mathbf{R}_\mathbf{v} \mathbf{q}_l} \\ &= \text{iSNR} \cdot \upsilon_{\text{sd}}(h_:) + \frac{1}{\xi_{\text{nr}}(h_:)}, \end{aligned} \tag{5.29}$$

where

$$\upsilon_{\text{sd}}(h_:) = \frac{J_x(h_:)}{\sum_{l=1}^{L} \phi_{c_{x,l}}}, \tag{5.30}$$

$$\xi_{\text{nr}}(h_:) = \frac{\sum_{l=1}^{L} \mathbf{q}_l^T \mathbf{R}_\mathbf{v} \mathbf{q}_l}{J_v(h_:)}. \tag{5.31}$$

Again, the fullband NMSE with the KLE depends explicitly on the fullband speech-distortion index and the fullband noise-reduction factor.

It is straightforward to see that minimizing the subband MSE for each l is equivalent to minimizing the fullband MSE.

5.3 WIENER FILTER

By minimizing $J(h_l)$ [eq. (5.22)] with respect to h_l, we easily find the Wiener gain:

$$\begin{aligned} h_{\mathrm{W},l} &= \frac{E\left[c_{x,l}^2(m)\right]}{E\left[c_{y,l}^2(m)\right]} \\ &= 1 - \frac{E\left[c_{v,l}^2(m)\right]}{E\left[c_{y,l}^2(m)\right]} \\ &= \frac{\phi_{c_{x,l}}}{\phi_{c_{x,l}} + \phi_{c_{v,l}}} \\ &= \frac{\mathrm{iSNR}_l}{1+\mathrm{iSNR}_l}. \end{aligned} \tag{5.32}$$

This gain is the equivalent form of the frequency-domain Wiener gain [7]. Clearly, $0 \leq h_{\mathrm{W},l} \leq 1,\ \forall l$.

We deduce the different subband performance measures:

$$\tilde{J}\left(h_{\mathrm{W},l}\right) = \frac{\mathrm{iSNR}_l}{1+\mathrm{iSNR}_l} \leq 1, \tag{5.33}$$

$$\begin{aligned} \xi_{\mathrm{nr}}(h_{\mathrm{W},l}) &= \xi_{\mathrm{sr}}(h_{\mathrm{W},l}) \\ &= \left(1+\frac{1}{\mathrm{iSNR}_l}\right)^2 \geq 1, \end{aligned} \tag{5.34}$$

$$\upsilon_{\mathrm{sd}}(h_{\mathrm{W},l}) = \frac{1}{(1+\mathrm{iSNR}_l)^2} \leq 1. \tag{5.35}$$

The fullband output SNR is

$$\mathrm{oSNR}(h_{\mathrm{W},:}) = \frac{\sum_{l=1}^{L} \phi_{c_{x,l}} \left(\frac{\mathrm{iSNR}_l}{1+\mathrm{iSNR}_l}\right)^2}{\sum_{l=1}^{L} \phi_{c_{v,l}} \left(\frac{\mathrm{iSNR}_l}{1+\mathrm{iSNR}_l}\right)^2}. \tag{5.36}$$

Property 5.1 With the optimal KLE-domain Wiener gain given in (5.32), the fullband output SNR is always greater than or equal to the input SNR, i.e., $\mathrm{oSNR}(h_{\mathrm{W},:}) \geq \mathrm{iSNR}$.

Proof. We can use exactly the same techniques as the ones exposed in [7] to show this property.
□

Property 5.2 We have

$$\frac{\text{iSNR}}{1+\text{oSNR}(h_{\text{W},:})} \leq \tilde{J}\left(h_{\text{W},:}\right) \leq \frac{\text{iSNR}}{1+\text{iSNR}}, \tag{5.37}$$

$$\frac{\left[1+\text{oSNR}(h_{\text{W},:})\right]^2}{\text{iSNR}\cdot\text{oSNR}(h_{\text{W},:})} \leq \xi_{\text{nr}}\left(h_{\text{W},:}\right) \leq \frac{(1+\text{iSNR})\left[1+\text{oSNR}(h_{\text{W},:})\right]}{\text{iSNR}^2}, \tag{5.38}$$

$$\frac{1}{\left[1+\text{oSNR}(h_{\text{W},:})\right]^2} \leq \upsilon_{\text{sd}}\left(h_{\text{W},:}\right) \leq \frac{1+\text{oSNR}(h_{\text{W},:})-\text{iSNR}}{(1+\text{iSNR})\left[1+\text{oSNR}(h_{\text{W},:})\right]}. \tag{5.39}$$

Proof. We can use exactly the same techniques as the ones exposed in [7] to show these different inequalities. □

It is of great interest to understand how the time-domain Wiener filter (see Chapter 3)

$$\mathbf{H}_{\text{t,W}} = \mathbf{Q}\left(\boldsymbol{\Lambda} - \mathbf{Q}^T\mathbf{R}_{\mathbf{v}}\mathbf{Q}\right)\boldsymbol{\Lambda}^{-1}\mathbf{Q}^T \tag{5.40}$$

is related to the KLE-domain Wiener gain given in (5.32).

Substituting the KLE-domain Wiener gain into (4.6), we see that the estimator of the vector $\mathbf{x}(m)$ can be written as

$$\begin{aligned} \mathbf{z}_{1,\text{W}}(m) &= \sum_{l=1}^{L} h_{\text{W},l}\, c_{y,l}(m)\mathbf{q}_l \\ &= \left(\sum_{l=1}^{L} h_{\text{W},l}\mathbf{q}_l\mathbf{q}_l^T\right)\mathbf{y}(m) \\ &= \mathbf{H}_{\text{TD,W}}\mathbf{y}(m). \end{aligned} \tag{5.41}$$

Therefore, the time-domain filtering matrix

$$\mathbf{H}_{\text{TD,W}} = \sum_{l=1}^{L} h_{\text{W},l}\mathbf{q}_l\mathbf{q}_l^T \tag{5.42}$$

is strictly equivalent to the KLE-domain gains $h_{\text{W},l},\ l = 1, 2, \ldots, L$. Substituting (5.32) into (5.42), we easily find that

$$\mathbf{H}_{\text{TD,W}} = \mathbf{Q}\left[\boldsymbol{\Lambda} - \text{diag}\left(\mathbf{Q}^T\mathbf{R}_{\mathbf{v}}\mathbf{Q}\right)\right]\boldsymbol{\Lambda}^{-1}\mathbf{Q}^T. \tag{5.43}$$

Clearly, the two filters $\mathbf{H}_{\text{t,W}}$ and $\mathbf{H}_{\text{TD,W}}$ may be very close to each other. For example if the noise is white, then $\mathbf{H}_{\text{t,W}} = \mathbf{H}_{\text{TD,W}}$. Also the orthogonal matrix $\mathbf{Q}$ tends to diagonalize the Toeplitz matrix $\mathbf{R}_{\mathbf{v}}$ for a large L. In this case, $\mathbf{Q}^T\mathbf{R}_{\mathbf{v}}\mathbf{Q} \approx \text{diag}\left(\mathbf{Q}^T\mathbf{R}_{\mathbf{v}}\mathbf{Q}\right)$, and as a result, $\mathbf{H}_{\text{t,W}} \approx \mathbf{H}_{\text{TD,W}}$.

5.4 TRADEOFF FILTER

The tradeoff gain is obtained by minimizing the speech distortion with the constraint that the residual noise level is equal to a value smaller than the level of the original noise. This is equivalent to solving the problem

$$\min_{h_l} J_x(h_l) \quad \text{subject to} \quad J_v(h_l) = \beta \phi_{c_{v,l}}, \tag{5.44}$$

where

$$\begin{aligned} J_x(h_l) &= (1 - h_l)^2 \phi_{c_{x,l}}, \qquad (5.45) \\ J_v(h_l) &= h_l^2 \phi_{c_{v,l}}, \qquad (5.46) \end{aligned}$$

and $0 < \beta < 1$ in order to have some noise reduction in the subband l. If we use a Lagrange multiplier, $\mu \geq 0$, to adjoin the constraint to the cost function, we get the tradeoff gain:

$$\begin{aligned} h_{\mathrm{T},\mu,l} &= \frac{\phi_{c_{x,l}}}{\phi_{c_{x,l}} + \mu \phi_{c_{v,l}}} \\ &= \frac{\lambda_l - \phi_{c_{v,l}}}{\lambda_l + (\mu - 1)\phi_{c_{v,l}}} \\ &= \frac{\mathrm{iSNR}_l}{\mu + \mathrm{iSNR}_l}. \end{aligned} \tag{5.47}$$

This gain can be seen as a KLE-domain Wiener gain with adjustable input noise level $\mu \phi_{c_{v,l}}$. The particular cases of $\mu = 1$ and $\mu = 0$ correspond to the Wiener and identity gains, respectively.

The fullband output SNR is

$$\mathrm{oSNR}(h_{\mathrm{T},\mu,:}) = \frac{\sum_{l=1}^{L} \phi_{c_{x,l}} \left(\frac{\mathrm{iSNR}_l}{\mu + \mathrm{iSNR}_l} \right)^2}{\sum_{l=1}^{L} \phi_{c_{v,l}} \left(\frac{\mathrm{iSNR}_l}{\mu + \mathrm{iSNR}_l} \right)^2}. \tag{5.48}$$

Property 5.3 With the tradeoff gain given in (5.47), the fullband output SNR is always greater than or equal to the input SNR, i.e., $\mathrm{oSNR}(h_{\mathrm{T},\mu,:}) \geq \mathrm{iSNR}, \ \forall \mu \geq 0$.

Proof. We can use exactly the same techniques as the ones exposed in [7] to show this property. □

From (5.48), we deduce that

$$\lim_{\mu \to \infty} \mathrm{oSNR}(h_{\mathrm{T},\mu,:}) = \frac{\sum_{l=1}^{L} \phi_{c_{x,l}} \mathrm{iSNR}_l^2}{\sum_{l=1}^{L} \phi_{c_{v,l}} \mathrm{iSNR}_l^2} \leq \sum_{l=1}^{L} \mathrm{iSNR}_l. \tag{5.49}$$

This shows how the fullband output SNR of the tradeoff gain is upper bounded.

The fullband speech-distortion index is

$$\upsilon_{\text{sd}}\left(h_{\text{T},\mu,:}\right) = \frac{\sum_{l=1}^{L} \frac{\phi_{c_{x,l}}\mu^2}{(\mu + \text{iSNR}_l)^2}}{\sum_{l=1}^{L} \phi_{c_{x,l}}}. \tag{5.50}$$

Property 5.4 The fullband speech-distortion index of the tradeoff gain is an increasing function of the parameter μ.

Proof. It is straightforward to verify that

$$\frac{d\upsilon_{\text{sd}}\left(h_{\text{T},\mu,:}\right)}{d\mu} \geq 0, \tag{5.51}$$

which ends the proof. □

It is clear that

$$0 \leq \upsilon_{\text{sd}}\left(h_{\text{T},\mu,:}\right) \leq 1, \ \forall \mu \geq 0. \tag{5.52}$$

Therefore, as μ increases, the fullband output SNR increases at the price of more distortion to the desired signal.

As we already did for the Wiener gain, we can write the KLE-domain tradeoff gain into the time domain. Indeed, substituting (5.47) into (4.6), we find that

$$\mathbf{H}_{\text{TD},\text{T},\mu} = \mathbf{Q}\left[\boldsymbol{\Lambda} - \text{diag}\left(\mathbf{Q}^T\mathbf{R}_{\mathbf{v}}\mathbf{Q}\right)\right]\left[\boldsymbol{\Lambda} + (\mu - 1) \cdot \text{diag}\left(\mathbf{Q}^T\mathbf{R}_{\mathbf{v}}\mathbf{Q}\right)\right]^{-1}\mathbf{Q}^T, \tag{5.53}$$

which has a similar form to the filtering matrix proposed in [46]. This matrix can be compared to the time-domain tradeoff filtering matrix (see Chapter 3)

$$\mathbf{H}_{\text{t},\text{T},\mu} = \mathbf{Q}\left(\boldsymbol{\Lambda} - \mathbf{Q}^T\mathbf{R}_{\mathbf{v}}\mathbf{Q}\right)\left[\boldsymbol{\Lambda} + (\mu - 1) \cdot \mathbf{Q}^T\mathbf{R}_{\mathbf{v}}\mathbf{Q}\right]^{-1}\mathbf{Q}^T. \tag{5.54}$$

We see that if the noise is white, the two matrices are the same.

5.5 MAXIMUM SNR FILTER

Let us define the $L \times 1$ vector

$$\mathbf{h} = \begin{bmatrix} h_1 & h_2 & \cdots & h_L \end{bmatrix}^T, \tag{5.55}$$

which contains all the subband gains. The fullband output SNR can be rewritten as

$$\begin{aligned} \text{oSNR}(h_:) &= \text{oSNR}(\mathbf{h}) \\ &= \frac{\mathbf{h}^T \mathbf{D}_{\phi_{c_x}} \mathbf{h}}{\mathbf{h}^T \mathbf{D}_{\phi_{c_v}} \mathbf{h}}, \end{aligned} \tag{5.56}$$

where

$$\mathbf{D}_{\phi_{c_x}} = \text{diag}\left(\phi_{c_{x,1}}, \phi_{c_{x,2}}, \ldots, \phi_{c_{x,L}}\right), \tag{5.57}$$

$$\mathbf{D}_{\phi_{c_v}} = \text{diag}\left(\phi_{c_{v,1}}, \phi_{c_{v,2}}, \ldots, \phi_{c_{v,L}}\right), \tag{5.58}$$

are two diagonal matrices. We assume here that $\phi_{c_{v,l}} \neq 0, \ \forall l$.

In the maximum SNR approach, we find the filter, $\mathbf{h}$, that maximizes the fullband output SNR defined in (5.56). The solution to this problem that we denote by $\mathbf{h}_{\text{max}}$ is simply the eigenvector corresponding to the maximum eigenvalue of the matrix $\mathbf{D}_{\phi_{c_v}}^{-1} \mathbf{D}_{\phi_{c_x}}$. Since this matrix is diagonal, its maximum eigenvalue is its largest diagonal element, i.e.,

$$\max_l \frac{\phi_{c_{x,l}}}{\phi_{c_{v,l}}} = \max_l \ \text{iSNR}_l. \tag{5.59}$$

Assume that this maximum is the l_0th diagonal element of the matrix $\mathbf{D}_{\phi_{c_v}}^{-1} \mathbf{D}_{\phi_{c_x}}$. In this case, the l_0th component of $\mathbf{h}_{\text{max}}$ is 1 and all its other components are 0. As a result,

$$\begin{aligned} \text{oSNR}(\mathbf{h}_{\text{max}}) &= \max_l \ \text{iSNR}_l \\ &= \text{iSNR}_{l_0}. \end{aligned} \tag{5.60}$$

We also deduce that

$$\text{oSNR}(h_:) \leq \max_l \ \text{iSNR}_l, \ \forall h_:. \tag{5.61}$$

This means that with the Wiener, tradeoff, or any other gain, the fullband output SNR cannot exceed the maximum subband input SNR, which is a very interesting result on its own.

It is easy to derive the fullband speech-distortion index:

$$\upsilon_{\text{sd}}\left(\mathbf{h}_{\text{max}}\right) = 1 - \frac{\phi_{c_{x,l_0}}}{\sum_{l=1}^{L} \phi_{c_{x,l}}}, \tag{5.62}$$

which can be very close to 1, implying very large distortions of the desired signal.

The equivalent time-domain version of $\mathbf{h}_{\text{max}}$ is simply

$$\mathbf{H}_{\text{TD,max}} = \mathbf{q}_{l_0} \mathbf{q}_{l_0}^T. \tag{5.63}$$

Needless to say that this maximum SNR filter is never used in practice since all subband signals but one are suppressed. But this filter is still interesting from a theoretical point of view.

CHAPTER 6

Optimal Filters in the KLE Domain with Model 2

In Model 2, the interframe correlation is taken into account. In this chapter, we show how to exploit this feature in order to develop noise reduction algorithms that are different from the ones developed with Model 1. To simplify the presentation, we drop the subscript "2" from the FIR filter of length M (see Chapter 4, Section 4.2), so that now $\mathbf{h}_{2,l}$ is written as $\mathbf{h}_l$.

6.1 PERFORMANCE MEASURES

From (4.16), we can deduce the most important performance measures.

The subband output SNR is defined as[1]

$$\begin{aligned} \mathrm{oSNR}(\mathbf{h}_l) &= \frac{\mathbf{h}_l^T \mathbf{\Phi}_{\mathbf{c}_{x_\mathrm{d}},l} \mathbf{h}_l}{\mathbf{h}_l^T \mathbf{\Phi}_{\mathrm{in},l} \mathbf{h}_l} \\ &= \frac{\phi_{c_{x,l}} \left(\mathbf{h}_l^T \boldsymbol{\gamma}_{c_{x,l}} \right)^2}{\mathbf{h}_l^T \mathbf{\Phi}_{\mathrm{in},l} \mathbf{h}_l}, \ l = 1, 2, \ldots, L, \end{aligned} \tag{6.1}$$

where

$$\mathbf{\Phi}_{\mathrm{in},l} = \mathbf{\Phi}_{\mathbf{c}'_{x,l}} + \mathbf{\Phi}_{\mathbf{c}_{v},l}, \ l = 1, 2, \ldots, L \tag{6.2}$$

is the interference-plus-noise correlation matrix. With Model 2, the subband output SNR is not equal, in general, to the subband input SNR contrary to Model 1. But for the particular filter $\mathbf{h}_l = \mathbf{i}_{M,1}$, where $\mathbf{i}_{M,1}$ is the first column of the identity matrix $\mathbf{I}_M$ of size $M \times M$, we have

$$\mathrm{oSNR}(\mathbf{i}_{M,1}) = \mathrm{iSNR}_l, \ l = 1, 2, \ldots, L. \tag{6.3}$$

For any two vectors $\mathbf{h}_l$ and $\boldsymbol{\gamma}_{c_{x,l}}$ and a positive definite matrix $\mathbf{\Phi}_{\mathrm{in},l}$, we have

$$\left(\mathbf{h}_l^T \boldsymbol{\gamma}_{c_{x,l}} \right)^2 \leq \left(\mathbf{h}_l^T \mathbf{\Phi}_{\mathrm{in},l} \mathbf{h}_l \right) \left(\boldsymbol{\gamma}_{c_{x,l}}^T \mathbf{\Phi}_{\mathrm{in},l}^{-1} \boldsymbol{\gamma}_{c_{x,l}} \right). \tag{6.4}$$

Using the previous inequality in (6.1), we deduce an upper bound for the subband output SNR:

$$\mathrm{oSNR}(\mathbf{h}_l) \leq \phi_{c_{x,l}} \boldsymbol{\gamma}_{c_{x,l}}^T \mathbf{\Phi}_{\mathrm{in},l}^{-1} \boldsymbol{\gamma}_{c_{x,l}}, \ l = 1, 2, \ldots, L. \tag{6.5}$$

[1]In this study, we consider the interference as part of the noise in the definitions of the performance measures.

We define the fullband output SNR as

$$\text{oSNR}(\mathbf{h}_{:}) = \frac{\sum_{l=1}^{L} \phi_{c_{x,l}} \left(\mathbf{h}_l^T \boldsymbol{\gamma}_{c_{x,l}} \right)^2}{\sum_{l=1}^{L} \mathbf{h}_l^T \boldsymbol{\Phi}_{\text{in},l} \mathbf{h}_l}. \tag{6.6}$$

We always have [7]

$$\text{oSNR}(\mathbf{h}_{:}) \leq \sum_{l=1}^{L} \text{oSNR}(\mathbf{h}_l) \leq \sum_{l=1}^{L} \phi_{c_{x,l}} \boldsymbol{\gamma}_{c_{x,l}}^T \boldsymbol{\Phi}_{\text{in},l}^{-1} \boldsymbol{\gamma}_{c_{x,l}}. \tag{6.7}$$

The previous inequality shows that the fullband output SNR is upper bounded no matter how the filters $\mathbf{h}_l$, $l = 1, 2, \ldots, L$ are chosen.

The subband and fullband noise-reduction factors are

$$\xi_{\text{nr}}(\mathbf{h}_l) = \frac{\phi_{c_{v,l}}}{\mathbf{h}_l^T \boldsymbol{\Phi}_{\text{in},l} \mathbf{h}_l}, \quad l = 1, 2, \ldots, L, \tag{6.8}$$

$$\xi_{\text{nr}}(\mathbf{h}_{:}) = \frac{\sum_{l=1}^{L} \phi_{c_{v,l}}}{\sum_{l=1}^{L} \mathbf{h}_l^T \boldsymbol{\Phi}_{\text{in},l} \mathbf{h}_l}. \tag{6.9}$$

These factors should be lower bounded by 1 for optimal filters. We also have

$$\xi_{\text{nr}}(\mathbf{h}_{:}) \leq \sum_{l=1}^{L} \xi_{\text{nr}}(\mathbf{h}_l). \tag{6.10}$$

From the inequality in (6.4), we easily find that

$$\xi_{\text{nr}}(\mathbf{h}_l) \leq \frac{\phi_{c_{v,l}} \boldsymbol{\gamma}_{c_{x,l}}^T \boldsymbol{\Phi}_{\text{in},l}^{-1} \boldsymbol{\gamma}_{c_{x,l}}}{\left(\mathbf{h}_l^T \boldsymbol{\gamma}_{c_{x,l}} \right)^2}, \quad l = 1, 2, \ldots, L, \tag{6.11}$$

$$\xi_{\text{nr}}(\mathbf{h}_{:}) \leq \sum_{l=1}^{L} \frac{\phi_{c_{v,l}} \boldsymbol{\gamma}_{c_{x,l}}^T \boldsymbol{\Phi}_{\text{in},l}^{-1} \boldsymbol{\gamma}_{c_{x,l}}}{\left(\mathbf{h}_l^T \boldsymbol{\gamma}_{c_{x,l}} \right)^2}. \tag{6.12}$$

To quantify the speech distortion, we give the subband speech-distortion index

$$\upsilon_{\text{sd}}(\mathbf{h}_l) = \frac{E\left\{ \left[c_{x,l}(m) \mathbf{h}_l^T \boldsymbol{\gamma}_{c_{x,l}} - c_{x,l}(m) \right]^2 \right\}}{\phi_{c_{x,l}}}$$

$$= \left(\mathbf{h}_l^T \boldsymbol{\gamma}_{c_{x,l}} - 1 \right)^2, \quad l = 1, 2, \ldots, L \tag{6.13}$$

and the fullband speech-distortion index

$$\begin{aligned} \upsilon_{\text{sd}}(\mathbf{h}_{:}) &= \frac{\sum_{l=1}^{L} \phi_{c_{x,l}} \left(\mathbf{h}_l^T \boldsymbol{\gamma}_{c_{x,l}} - 1\right)^2}{\sum_{l=1}^{L} \phi_{c_{x,l}}} \\ &= \frac{\sum_{l=1}^{L} \upsilon_{\text{sd}}(\mathbf{h}_l) \phi_{c_{x,l}}}{\sum_{l=1}^{L} \phi_{c_{x,l}}}. \end{aligned} \tag{6.14}$$

The speech-distortion index is usually upper bounded by 1. We have

$$\upsilon_{\text{sd}}(\mathbf{h}_{:}) \leq \sum_{l=1}^{L} \upsilon_{\text{sd}}(\mathbf{h}_l). \tag{6.15}$$

We can also quantify signal distortion via the subband and fullband speech-reduction factors which are defined as

$$\begin{aligned} \xi_{\text{sr}}(\mathbf{h}_l) &= \frac{\phi_{c_{x,l}}}{\phi_{c_{x,l}} \left(\mathbf{h}_l^T \boldsymbol{\gamma}_{c_{x,l}}\right)^2} \\ &= \frac{1}{\left(\mathbf{h}_l^T \boldsymbol{\gamma}_{c_{x,l}}\right)^2}, \ l = 1, 2, \ldots, L, \end{aligned} \tag{6.16}$$

$$\begin{aligned} \xi_{\text{sr}}(\mathbf{h}_{:}) &= \frac{\sum_{l=1}^{L} \phi_{c_{x,l}}}{\sum_{l=1}^{L} \phi_{c_{x,l}} \left(\mathbf{h}_l^T \boldsymbol{\gamma}_{c_{x,l}}\right)^2} \\ &= \frac{\sum_{l=1}^{L} \phi_{c_{x,l}}}{\sum_{l=1}^{L} \xi_{\text{sr}}^{-1}(\mathbf{h}_l) \phi_{c_{x,l}}}. \end{aligned} \tag{6.17}$$

The speech-reduction factor is supposed to have a lower bound of 1 for optimal filters. We also have

$$\xi_{\text{sr}}(\mathbf{h}_{:}) \leq \sum_{l=1}^{L} \xi_{\text{sr}}(\mathbf{h}_l). \tag{6.18}$$

A key observation from (6.13) or (6.16) is that the design of a noise reduction algorithm that does not distort the desired signal requires the constraint

$$\mathbf{h}_l^T \boldsymbol{\gamma}_{c_{x,l}} = 1, \ \forall l. \tag{6.19}$$

It can easily be checked that

$$\frac{\text{oSNR}(\mathbf{h}_l)}{\text{iSNR}_l} = \frac{\xi_{\text{nr}}(\mathbf{h}_l)}{\xi_{\text{sr}}(\mathbf{h}_l)}, \ l = 1, 2, \ldots, L, \tag{6.20}$$

$$\frac{\text{oSNR}(\mathbf{h}_{:})}{\text{iSNR}} = \frac{\xi_{\text{nr}}(\mathbf{h}_{:})}{\xi_{\text{sr}}(\mathbf{h}_{:})}. \tag{6.21}$$

6.2 MAXIMUM SNR FILTER

The maximum SNR filter, $\mathbf{h}_{\max,l}$, is obtained by maximizing the subband output SNR as defined in (6.1). Therefore, $\mathbf{h}_{\max,l}$ is the eigenvector corresponding to the maximum eigenvalue of the matrix $\mathbf{\Phi}_{\text{in},l}^{-1}\mathbf{\Phi}_{\mathbf{c}_{x_\text{d}},l}$. Let us denote this eigenvalue by $\lambda_{\max,l}$. Since the rank of the matrix $\mathbf{\Phi}_{\mathbf{c}_{x_\text{d}},l}$ is equal to 1, we have

$$\begin{aligned}\lambda_{\max,l} &= \text{tr}\left(\mathbf{\Phi}_{\text{in},l}^{-1}\mathbf{\Phi}_{\mathbf{c}_{x_\text{d}},l}\right) \\ &= \phi_{c_{x,l}}\boldsymbol{\gamma}_{c_{x,l}}^T\mathbf{\Phi}_{\text{in},l}^{-1}\boldsymbol{\gamma}_{c_{x,l}},\ l = 1, 2, \ldots, L.\end{aligned} \tag{6.22}$$

As a result,

$$\text{oSNR}(\mathbf{h}_{\max,l}) = \phi_{c_{x,l}}\boldsymbol{\gamma}_{c_{x,l}}^T\mathbf{\Phi}_{\text{in},l}^{-1}\boldsymbol{\gamma}_{c_{x,l}},\ l = 1, 2, \ldots, L, \tag{6.23}$$

which corresponds to the maximum possible output SNR according to the inequality in (6.5). Obviously, we also have

$$\mathbf{h}_{\max,l} = \alpha_l\mathbf{\Phi}_{\text{in},l}^{-1}\boldsymbol{\gamma}_{c_{x,l}},\ l = 1, 2, \ldots, L, \tag{6.24}$$

where α_l is an arbitrary scaling factor different from zero. While this factor has no effect on the subband output SNR, it has on the fullband output SNR and speech distortion (subband and fullband). In fact, all filters derived in the rest of this chapter are equivalent up to this scaling factor. These filters also try to find the respective scaling factors depending on what we optimize.

6.3 MSE CRITERION

The error signal between the estimated and desired signals in the subband l is

$$\begin{aligned}e_l(m) &= c_{z_2,l}(m) - c_{x,l}(m) \\ &= \mathbf{h}_l^T\mathbf{c}_{y,l}(m) - c_{x,l}(m).\end{aligned} \tag{6.25}$$

This error signal can also be written as the sum of two uncorrelated error signals:

$$e_l(m) = e_{x,l}(m) + e_{\text{in},l}(m), \tag{6.26}$$

where

$$\begin{aligned}e_{x,l}(m) &= \mathbf{h}_l^T\mathbf{c}_{x_\text{d},l}(m) - c_{x,l}(m) \\ &= \left(\mathbf{h}_l^T\boldsymbol{\gamma}_{c_{x,l}} - 1\right)c_{x,l}(m)\end{aligned} \tag{6.27}$$

is the speech distortion due to the filter and

$$e_{\text{in},l}(m) = \mathbf{h}_l^T\mathbf{c}'_{x,l}(m) + \mathbf{h}_l^T\mathbf{c}_{v,l}(m) \tag{6.28}$$

represents the residual interference-plus-noise.

The subband MSE criterion is then

$$
\begin{aligned}
J(\mathbf{h}_l) &= E\left[e_l^2(m)\right] \\
&= \mathbf{h}_l^T \mathbf{\Phi}_{\mathbf{c}_{y,l}} \mathbf{h}_l - 2\mathbf{h}_l^T \mathbf{\Phi}_{\mathbf{c}_{y,l}\mathbf{c}_{x,l}} \mathbf{i}_{M,1} + \phi_{c_{x,l}},
\end{aligned}
\tag{6.29}
$$

where

$$
\begin{aligned}
\mathbf{\Phi}_{\mathbf{c}_{y,l}\mathbf{c}_{x,l}} &= E\left[\mathbf{c}_{y,l}(m)\mathbf{c}_{x,l}^T(m)\right] \\
&= E\left[\mathbf{c}_{x,l}(m)\mathbf{c}_{x,l}^T(m)\right] \\
&= \mathbf{\Phi}_{\mathbf{c}_{x,l}}
\end{aligned}
$$

is the cross-correlation matrix between the two signal vectors $\mathbf{c}_{y,l}(m)$ and $\mathbf{c}_{x,l}(m)$. We can rewrite the subband MSE as

$$
J(\mathbf{h}_l) = J_x(\mathbf{h}_l) + J_{\text{in}}(\mathbf{h}_l),
$$

where

$$
\begin{aligned}
J_x(\mathbf{h}_l) &= E\left[e_{x,l}^2(m)\right] \\
&= \phi_{c_{x,l}} \left(\mathbf{h}_l^T \boldsymbol{\gamma}_{c_{x,l}} - 1\right)^2
\end{aligned}
\tag{6.30}
$$

and

$$
\begin{aligned}
J_{\text{in}}(\mathbf{h}_l) &= E\left[e_{\text{in},l}^2(m)\right] \\
&= \mathbf{h}_l^T \mathbf{\Phi}_{\text{in},l} \mathbf{h}_l.
\end{aligned}
\tag{6.31}
$$

For the particular filter $\mathbf{h}_l = \mathbf{i}_{M,1},\ \forall l$, we get

$$
J(\mathbf{i}_{M,1}) = \phi_{c_{v,l}}.
\tag{6.32}
$$

Using this particular case of the MSE, we define the subband normalized MSE (NMSE) as

$$
\begin{aligned}
\tilde{J}(\mathbf{h}_l) &= \frac{J(\mathbf{h}_l)}{J(\mathbf{i}_{M,1})} \\
&= \text{iSNR}_l \cdot \upsilon_{\text{sd}}(\mathbf{h}_l) + \frac{1}{\xi_{\text{nr}}(\mathbf{h}_l)},
\end{aligned}
\tag{6.33}
$$

where

$$
\upsilon_{\text{sd}}(\mathbf{h}_l) = \frac{J_x(\mathbf{h}_l)}{\phi_{c_{x,l}}},
\tag{6.34}
$$

$$
\xi_{\text{nr}}(\mathbf{h}_l) = \frac{\phi_{c_{v,l}}}{J_{\text{in}}(\mathbf{h}_l)}.
\tag{6.35}
$$

The KLE-domain NMSE depends explicitly on the subband speech-distortion index and the subband noise-reduction factor.

We define the fullband MSE and fullband NMSE as

$$\begin{aligned} J\left(\mathbf{h}_{:}\right) &= \frac{1}{L}\sum_{l=1}^{L} J\left(\mathbf{h}_l\right) \\ &= \frac{1}{L}\sum_{l=1}^{L} J_x\left(\mathbf{h}_l\right) + \frac{1}{L}\sum_{l=1}^{L} J_{\text{in}}\left(\mathbf{h}_l\right) \\ &= J_x\left(\mathbf{h}_{:}\right) + J_{\text{in}}\left(\mathbf{h}_{:}\right) \end{aligned} \tag{6.36}$$

and

$$\begin{aligned} \tilde{J}\left(\mathbf{h}_{:}\right) &= L\frac{J\left(\mathbf{h}_{:}\right)}{\sum_{l=1}^{L}\phi_{c_{v,l}}} \\ &= \text{iSNR}\cdot\upsilon_{\text{sd}}(\mathbf{h}_{:}) + \frac{1}{\xi_{\text{nr}}(\mathbf{h}_{:})}, \end{aligned} \tag{6.37}$$

where

$$\upsilon_{\text{sd}}(\mathbf{h}_{:}) = \frac{J_x\left(\mathbf{h}_{:}\right)}{\sum_{l=1}^{L}\phi_{c_{x,l}}}, \tag{6.38}$$

$$\xi_{\text{nr}}(\mathbf{h}_{:}) = \frac{\sum_{l=1}^{L}\phi_{c_{v,l}}}{J_{\text{in}}\left(\mathbf{h}_{:}\right)}. \tag{6.39}$$

The fullband NMSE with the KLE depends also explicitly on the fullband speech-distortion index and the fullband noise-reduction factor.

It is straightforward to see that minimizing the subband MSE for each l is equivalent to minimizing the fullband MSE.

6.4 WIENER FILTER

The Wiener filter is easily derived by taking the gradient of the MSE, $J\left(\mathbf{h}_l\right)$, with respect to $\mathbf{h}_l$ and equating the result to zero:

$$\begin{aligned} \mathbf{h}_{\text{W},l} &= \mathbf{\Phi}_{\mathbf{c}_{y,l}}^{-1}\mathbf{\Phi}_{\mathbf{c}_{x,l}}\mathbf{i}_{M,1} \\ &= \left(\mathbf{I}_M - \mathbf{\Phi}_{\mathbf{c}_{y,l}}^{-1}\mathbf{\Phi}_{\mathbf{c}_{v,l}}\right)\mathbf{i}_{M,1}. \end{aligned} \tag{6.40}$$

Since

$$\mathbf{\Phi}_{\mathbf{c}_{x,l}}\mathbf{i}_{M,1} = \phi_{c_{x,l}}\boldsymbol{\gamma}_{c_{x,l}}, \tag{6.41}$$

we can rewrite (6.40) as

$$\mathbf{h}_{\mathrm{W},l} = \phi_{c_{x,l}} \mathbf{\Phi}_{\mathbf{c}_{y,l}}^{-1} \boldsymbol{\gamma}_{c_{x,l}}. \tag{6.42}$$

It is easy to verify that

$$\mathbf{\Phi}_{\mathbf{c}_{y,l}} = \phi_{c_{x,l}} \boldsymbol{\gamma}_{c_{x,l}} \boldsymbol{\gamma}_{c_{x,l}}^T + \mathbf{\Phi}_{\mathrm{in},l}. \tag{6.43}$$

Determining the inverse of $\mathbf{\Phi}_{\mathbf{c}_{y,l}}$ from (6.43) with the Woodbury's identity

$$\mathbf{\Phi}_{\mathbf{c}_{y,l}}^{-1} = \mathbf{\Phi}_{\mathrm{in},l}^{-1} - \frac{\mathbf{\Phi}_{\mathrm{in},l}^{-1} \boldsymbol{\gamma}_{c_{x,l}} \boldsymbol{\gamma}_{c_{x,l}}^T \mathbf{\Phi}_{\mathrm{in},l}^{-1}}{\phi_{c_{x,l}}^{-1} + \boldsymbol{\gamma}_{c_{x,l}}^T \mathbf{\Phi}_{\mathrm{in},l}^{-1} \boldsymbol{\gamma}_{c_{x,l}}} \tag{6.44}$$

and substituting the result into (6.42), leads to another interesting formulation of the Wiener filter:

$$\mathbf{h}_{\mathrm{W},l} = \frac{\mathbf{\Phi}_{\mathrm{in},l}^{-1} \boldsymbol{\gamma}_{c_{x,l}}}{\phi_{c_{x,l}}^{-1} + \boldsymbol{\gamma}_{c_{x,l}}^T \mathbf{\Phi}_{\mathrm{in},l}^{-1} \boldsymbol{\gamma}_{c_{x,l}}}, \tag{6.45}$$

that we can rewrite as

$$\begin{aligned} \mathbf{h}_{\mathrm{W},l} &= \frac{\mathbf{\Phi}_{\mathrm{in},l}^{-1} \mathbf{\Phi}_{\mathbf{c}_{y,l}} - \mathbf{I}_M}{1 - M + \mathrm{tr}\left(\mathbf{\Phi}_{\mathrm{in},l}^{-1} \mathbf{\Phi}_{\mathbf{c}_{y,l}}\right)} \mathbf{i}_{M,1} \\ &= \frac{\mathbf{\Phi}_{\mathrm{in},l}^{-1} \mathbf{\Phi}_{\mathbf{c}_{x_{\mathrm{d}},l}}}{1 + \lambda_{\max,l}} \mathbf{i}_{M,1}. \end{aligned} \tag{6.46}$$

We can deduce from (6.45) that the subband output SNR is

$$\begin{aligned} \mathrm{oSNR}\left(\mathbf{h}_{\mathrm{W},l}\right) &= \lambda_{\max,l} \\ &= \mathrm{tr}\left(\mathbf{\Phi}_{\mathrm{in},l}^{-1} \mathbf{\Phi}_{\mathbf{c}_{y,l}}\right) - M, \end{aligned} \tag{6.47}$$

and the subband speech-distortion index is a clear function of the subband output SNR:

$$\upsilon_{\mathrm{sd}}\left(\mathbf{h}_{\mathrm{W},l}\right) = \frac{1}{\left[1 + \mathrm{oSNR}\left(\mathbf{h}_{\mathrm{W},l}\right)\right]^2}. \tag{6.48}$$

The higher is the value of $\mathrm{oSNR}\left(\mathbf{h}_{\mathrm{W},l}\right)$, the less the desired signal is distorted.

Clearly,

$$\mathrm{oSNR}\left(\mathbf{h}_{\mathrm{W},l}\right) \geq \mathrm{iSNR}_l, \tag{6.49}$$

since the Wiener filter maximizes the subband output SNR. Recall that in Model 1, the subband output SNR cannot be improved.

It is of great interest to observe that the two filters, $\mathbf{h}_{\max,l}$ and $\mathbf{h}_{\mathrm{W},l}$ are equivalent up to a scaling factor. Indeed, taking

$$\alpha_l = \frac{\phi_{c_{x,l}}}{1 + \lambda_{\max,l}} \tag{6.50}$$

in (6.24) (maximum SNR filter), we find (6.46) (Wiener filter).

With the Wiener filter, the subband noise-reduction factor is

$$\begin{aligned} \xi_{\mathrm{nr}}\left(\mathbf{h}_{\mathrm{W},l}\right) &= \frac{\left[1 + \mathrm{oSNR}\left(\mathbf{h}_{\mathrm{W},l}\right)\right]^2}{\mathrm{iSNR}_l \cdot \mathrm{oSNR}\left(\mathbf{h}_{\mathrm{W},l}\right)} \\ &\geq \left[1 + \frac{1}{\mathrm{oSNR}\left(\mathbf{h}_{\mathrm{W},l}\right)}\right]^2 . \end{aligned} \tag{6.51}$$

Using (6.48) and (6.51) in (6.33), we find the minimum NMSE:

$$\tilde{J}\left(\mathbf{h}_{\mathrm{W},l}\right) = \frac{\mathrm{iSNR}_l}{1 + \mathrm{oSNR}\left(\mathbf{h}_{\mathrm{W},l}\right)} \leq 1. \tag{6.52}$$

The fullband output SNR is

$$\mathrm{oSNR}(\mathbf{h}_{\mathrm{W},:}) = \frac{\sum_{l=1}^{L} \phi_{c_{x,l}} \frac{\mathrm{oSNR}^2\left(\mathbf{h}_{\mathrm{W},l}\right)}{\left[1 + \mathrm{oSNR}\left(\mathbf{h}_{\mathrm{W},l}\right)\right]^2}}{\sum_{l=1}^{L} \phi_{c_{x,l}} \frac{\mathrm{oSNR}\left(\mathbf{h}_{\mathrm{W},l}\right)}{\left[1 + \mathrm{oSNR}\left(\mathbf{h}_{\mathrm{W},l}\right)\right]^2}} . \tag{6.53}$$

Property 6.1 With the optimal KLE-domain Wiener filter given in (6.40), the fullband output SNR is always greater than or equal to the input SNR, i.e., $\mathrm{oSNR}(\mathbf{h}_{\mathrm{W},:}) \geq \mathrm{iSNR}$.

Proof. We can use exactly the same techniques as the ones exposed in [7] to show this property. □

6.5 MINIMUM VARIANCE DISTORTIONLESS RESPONSE (MVDR) FILTER

The celebrated minimum variance distortionless response (MVDR) filter proposed by Capon [10], [36] is usually derived in a context where we have at least two sensors (or microphones) available. Interestingly, with Model 2, we can also derive the MVDR (with one sensor only) by minimizing the MSE of the residual interference-plus-noise, $J_{\text{in}}\left(\mathbf{h}_l\right)$, with the constraint that the desired signal is not distorted. Mathematically, this is equivalent to

$$\min_{\mathbf{h}_l} \mathbf{h}_l^T \mathbf{\Phi}_{\text{in},l} \mathbf{h}_l \quad \text{subject to} \quad \mathbf{h}_l^T \boldsymbol{\gamma}_{c_{x,l}} = 1, \tag{6.54}$$

for which the solution is

$$\begin{aligned} \mathbf{h}_{\text{MVDR},l} &= \frac{\phi_{c_{x,l}} \mathbf{\Phi}_{\text{in},l}^{-1} \boldsymbol{\gamma}_{c_{x,l}}}{\lambda_{\max,l}} \\ &= \frac{\mathbf{\Phi}_{\text{in},l}^{-1} \mathbf{\Phi}_{\mathbf{c}_{y,l}} - \mathbf{I}_M}{\text{tr}\left(\mathbf{\Phi}_{\text{in},l}^{-1} \mathbf{\Phi}_{\mathbf{c}_{y,l}}\right) - M} \mathbf{i}_{M,1}. \end{aligned} \tag{6.55}$$

Obviously, we can rewrite the MVDR as

$$\mathbf{h}_{\text{MVDR},l} = \frac{\mathbf{\Phi}_{\mathbf{c}_{y,l}}^{-1} \boldsymbol{\gamma}_{c_{x,l}}}{\boldsymbol{\gamma}_{c_{x,l}}^T \mathbf{\Phi}_{\mathbf{c}_{y,l}}^{-1} \boldsymbol{\gamma}_{c_{x,l}}}. \tag{6.56}$$

Taking

$$\alpha_l = \frac{\phi_{c_{x,l}}}{\lambda_{\max,l}} \tag{6.57}$$

in (6.24) (maximum SNR filter), we find (6.55) (MVDR filter), showing how the maximum SNR, MVDR, and Wiener filters are equivalent up to a scaling factor. From a subband point of view, this scaling is not significant, but from a fullband point of view, it can be important since speech signals are broadband in nature. Indeed, it can easily be verified that this scaling factor affects the fullband output SNRs and fullband speech-distortion indices. While the subband output SNRs of the maximum SNR, Wiener, and MVDR filters are the same, the fullband output SNRs are not because of the scaling factor.

It is clear that we always have

$$\text{oSNR}\left(\mathbf{h}_{\text{MVDR},l}\right) = \text{oSNR}\left(\mathbf{h}_{\text{W},l}\right), \tag{6.58}$$

$$\upsilon_{\text{sd}}\left(\mathbf{h}_{\text{MVDR},l}\right) = 0, \tag{6.59}$$

$$\xi_{\text{sr}}\left(\mathbf{h}_{\text{MVDR},l}\right) = 1, \tag{6.60}$$

$$\xi_{\text{nr}}\left(\mathbf{h}_{\text{MVDR},l}\right) = \frac{\lambda_{\max,l}}{\text{iSNR}_l} \leq \xi_{\text{nr}}\left(\mathbf{h}_{\text{W},l}\right), \tag{6.61}$$

and

$$1 \geq \tilde{J}\left(\mathbf{h}_{\text{MVDR},l}\right) = \frac{\text{iSNR}_l}{\lambda_{\max,l}} \geq \tilde{J}\left(\mathbf{h}_{\text{W},l}\right). \tag{6.62}$$

The fullband output SNR is

$$\text{oSNR}(\mathbf{h}_{\text{MVDR},:}) = \frac{\sum_{l=1}^{L} \phi_{c_{x,l}}}{\sum_{l=1}^{L} \frac{\phi_{c_{x,l}}}{\text{oSNR}\left(\mathbf{h}_{\text{MVDR},l}\right)}}. \tag{6.63}$$

Property 6.2 With the optimal KLE-domain MVDR filter given in (6.55), the fullband output SNR is always greater than or equal to the input SNR, i.e., $\text{oSNR}(\mathbf{h}_{\text{MVDR},:}) \geq \text{iSNR}$.

Proof. See next section. □

6.6 TRADEOFF FILTER

In the tradeoff approach, we try to compromise between noise reduction and speech distortion. Instead of minimizing the MSE to find the Wiener filter or minimizing the MSE of the residual interference-plus-noise with the constraint of no distortion to find the MVDR, we could minimize the speech-distortion index with the constraint that the noise-reduction factor is equal to a positive value that is greater than 1. Mathematically, this is equivalent to

$$\min_{\mathbf{h}_l} J_x\left(\mathbf{h}_l\right) \quad \text{subject to} \quad J_{\text{in}}\left(\mathbf{h}_l\right) = \beta \phi_{c_{v,l}}, \tag{6.64}$$

where $0 < \beta < 1$ to insure that we get some noise reduction. By using a Lagrange multiplier, $\mu > 0$, to adjoin the constraint to the cost function, we easily deduce the tradeoff filter:

$$\begin{aligned} \mathbf{h}_{\text{T},\mu,l} &= \phi_{c_{x,l}} \left(\phi_{c_{x,l}} \boldsymbol{\gamma}_{c_{x,l}} \boldsymbol{\gamma}_{c_{x,l}}^T + \mu \boldsymbol{\Phi}_{\text{in},l} \right)^{-1} \boldsymbol{\gamma}_{c_{x,l}} \\ &= \frac{\phi_{c_{x,l}} \boldsymbol{\Phi}_{\text{in},l}^{-1} \boldsymbol{\gamma}_{c_{x,l}}}{\mu + \lambda_{\max,l}}, \end{aligned} \tag{6.65}$$

where the Lagrange multiplier, μ, satisfies $J_{\text{in}}\left(\mathbf{h}_{\text{T},\mu,l}\right) = \beta \phi_{c_{v,l}}$. However, in practice, it is not easy to determine the optimal μ. Therefore, when this parameter is chosen in an ad-hoc way, we can see that for

- $\mu = 1$, $\mathbf{h}_{\text{T},1,l} = \mathbf{h}_{\text{W},l}$, which is the Wiener filter;
- $\mu = 0$, $\mathbf{h}_{\text{T},0,l} = \mathbf{h}_{\text{MVDR},l}$, which is the MVDR filter;
- $\mu > 1$, results in low residual noise at the expense of high speech distortion;
- $\mu < 1$, results in high residual noise and low speech distortion.

Note that the MVDR filter cannot be derived from the first line of (6.65) since by taking $\mu = 0$, we have to invert a matrix that is not full rank.

Again, we observe here as well that the tradeoff and Wiener filters are equivalent up to a scaling factor. As a result, the subband output SNR with the tradeoff filter is obviously the same as the subband output SNR with the Wiener filter, i.e.,

$$\text{oSNR}\left(\mathbf{h}_{\text{T},\mu,l}\right) = \lambda_{\max,l}, \tag{6.66}$$

and does not depend on μ. However, the subband speech-distortion index is now both a function of the variable μ and the subband output SNR:

$$\upsilon_{\text{sd}}\left(\mathbf{h}_{\text{T},\mu,l}\right) = \frac{\mu^2}{\left(\mu + \lambda_{\max,l}\right)^2}. \tag{6.67}$$

From (6.67), we observe how μ can affect the desired signal.

The tradeoff filter is interesting from several perspectives since it encompasses both the Wiener and MVDR filters. It is then useful to study the fullband output SNR and the fullband speech-distortion index of the tradeoff filter, which both depend on the variable μ.

Using (6.65) in (6.6), we find that the fullband output SNR is

$$\text{oSNR}\left(\mathbf{h}_{\text{T},\mu,:}\right) = \frac{\sum_{l=1}^{L} \frac{\phi_{C_X,l}\lambda_{\max,l}^2}{\left(\mu + \lambda_{\max,l}\right)^2}}{\sum_{l=1}^{L} \frac{\phi_{C_X,l}\lambda_{\max,l}}{\left(\mu + \lambda_{\max,l}\right)^2}}. \tag{6.68}$$

We propose the following.

Property 6.3 The fullband output SNR of the tradeoff filter is an increasing function of the parameter μ.

Proof. The proof is very similar to the one given in [49].

In order to determine the variations of $\text{oSNR}\left(\mathbf{h}_{\text{T},\mu,:}\right)$ with respect to the parameter μ, we will check the sign of the following differentiation with respect to μ:

$$\frac{d\text{oSNR}\left(\mathbf{h}_{\text{T},\mu,:}\right)}{d\mu} = 2\frac{\text{Num}(\mu)}{\text{Den}(\mu)}, \tag{6.69}$$

where

$$\begin{aligned} \text{Num}(\mu) &= -\sum_{l=1}^{L} \frac{\phi_{C_X,l}\lambda_{\max,l}}{\left(\mu + \lambda_{\max,l}\right)^2} \sum_{l=1}^{L} \frac{\phi_{C_X,l}\lambda_{\max,l}^2}{\left(\mu + \lambda_{\max,l}\right)^3} \\ &\quad + \sum_{l=1}^{L} \frac{\phi_{C_X,l}\lambda_{\max,l}^2}{\left(\mu + \lambda_{\max,l}\right)^2} \sum_{l=1}^{L} \frac{\phi_{C_X,l}\lambda_{\max,l}}{\left(\mu + \lambda_{\max,l}\right)^3} \end{aligned} \tag{6.70}$$

and

$$\mathrm{Den}(\mu) = \left[\sum_{l=1}^{L} \frac{\phi_{C_{x,l}} \lambda_{\max,l}}{\left(\mu + \lambda_{\max,l}\right)^2} \right]^2 . \tag{6.71}$$

We only focus on the numerator of the above derivative to see the variations of the fullband output SNR since the denominator is always positive. Multiplying and dividing by $\mu + \lambda_{\max,l}$, this numerator can be rewritten as

$$\begin{aligned}
\mathrm{Num}(\mu) &= -\sum_{l=1}^{L} \frac{\phi_{C_{x,l}} \lambda_{\max,l} \left(\mu + \lambda_{\max,l}\right)}{\left(\mu + \lambda_{\max,l}\right)^3} \sum_{l=1}^{L} \frac{\phi_{C_{x,l}} \lambda_{\max,l}^2}{\left(\mu + \lambda_{\max,l}\right)^3} \\
&\quad + \sum_{l=1}^{L} \frac{\phi_{C_{x,l}} \lambda_{\max,l}^2 \left(\mu + \lambda_{\max,l}\right)}{\left(\mu + \lambda_{\max,l}\right)^3} \sum_{l=1}^{L} \frac{\phi_{C_{x,l}} \lambda_{\max,l}}{\left(\mu + \lambda_{\max,l}\right)^3} \\
&= -\left[\sum_{l=1}^{L} \frac{\phi_{C_{x,l}} \lambda_{\max,l}^2}{\left(\mu + \lambda_{\max,l}\right)^3} \right]^2 \\
&\quad - \mu \sum_{l=1}^{L} \frac{\phi_{C_{x,l}} \lambda_{\max,l}}{\left(\mu + \lambda_{\max,l}\right)^3} \sum_{l=1}^{L} \frac{\phi_{C_{x,l}} \lambda_{\max,l}^2}{\left(\mu + \lambda_{\max,l}\right)^3} \\
&\quad + \sum_{l=1}^{L} \frac{\phi_{C_{x,l}} \lambda_{\max,l}^3}{\left(\mu + \lambda_{\max,l}\right)^3} \sum_{l=1}^{L} \frac{\phi_{C_{x,l}} \lambda_{\max,l}}{\left(\mu + \lambda_{\max,l}\right)^3} \\
&\quad + \mu \sum_{l=1}^{L} \frac{\phi_{C_{x,l}} \lambda_{\max,l}}{\left(\mu + \lambda_{\max,l}\right)^3} \sum_{l=1}^{L} \frac{\phi_{C_{x,l}} \lambda_{\max,l}^2}{\left(\mu + \lambda_{\max,l}\right)^3} \\
&= -\left[\sum_{l=1}^{L} \frac{\phi_{C_{x,l}} \lambda_{\max,l}^2}{\left(\mu + \lambda_{\max,l}\right)^3} \right]^2 \\
&\quad + \sum_{l=1}^{L} \frac{\phi_{C_{x,l}} \lambda_{\max,l}^3}{\left(\mu + \lambda_{\max,l}\right)^3} \sum_{l=1}^{L} \frac{\phi_{C_{x,l}} \lambda_{\max,l}}{\left(\mu + \lambda_{\max,l}\right)^3} .
\end{aligned} \tag{6.72}$$

As far as μ, $\lambda_{\max,l}$, and $\phi_{C_{x,l}}$ are positive $\forall l$, we can use the Cauchy-Schwarz inequality

$$\begin{aligned}
& \sum_{l=1}^{L} \frac{\phi_{C_{x,l}} \lambda_{\max,l}^3}{\left(\mu + \lambda_{\max,l}\right)^3} \sum_{l=1}^{L} \frac{\phi_{C_{x,l}} \lambda_{\max,l}}{\left(\mu + \lambda_{\max,l}\right)^3} \\
& \geq \left[\sum_{l=1}^{L} \sqrt{\frac{\phi_{C_{x,l}} \lambda_{\max,l}^3}{\left(\mu + \lambda_{\max,l}\right)^3}} \sqrt{\frac{\phi_{C_{x,l}} \lambda_{\max,l}}{\left(\mu + \lambda_{\max,l}\right)^3}} \right]^2 \\
& = \left[\sum_{l=1}^{L} \frac{\phi_{C_{x,l}} \lambda_{\max,l}^2}{\left(\mu + \lambda_{\max,l}\right)^3} \right]^2 .
\end{aligned} \tag{6.73}$$

Substituting (6.73) into (6.72), we conclude that

$$\frac{d\text{oSNR}\left(\mathbf{h}_{\text{T},\mu,:}\right)}{d\mu} \geq 0, \tag{6.74}$$

proving that the fullband output SNR is increasing with respect to μ. □

From Property 6.3, we deduce that the MVDR filter gives the smallest fullband output SNR, which is

$$\text{oSNR}\left(\mathbf{h}_{\text{T},0,:}\right) = \frac{\sum_{l=1}^{L} \phi_{c_{x,l}}}{\sum_{l=1}^{L} \frac{\phi_{c_{x,l}}}{\lambda_{\max,l}}}. \tag{6.75}$$

We give another interesting property.

Property 6.4 We have

$$\lim_{\mu\to\infty} \text{oSNR}\left(\mathbf{h}_{\text{T},\mu,:}\right) = \frac{\sum_{l=1}^{L} \phi_{c_{x,l}} \lambda_{\max,l}^2}{\sum_{l=1}^{L} \phi_{c_{x,l}} \lambda_{\max,l}} \leq \sum_{l=1}^{L} \lambda_{\max,l}. \tag{6.76}$$

Proof. Easy to show from (6.68). □

While the fullband output SNR is upper bounded, it is easy to show that the fullband noise-reduction factor and fullband speech-reduction factor are not. So when μ goes to infinity, so are $\xi_{\text{nr}}\left(\mathbf{h}_{\text{T},\mu,:}\right)$ and $\xi_{\text{sr}}\left(\mathbf{h}_{\text{T},\mu,:}\right)$.

The fullband speech-distortion index is

$$\upsilon_{\text{sd}}\left(\mathbf{h}_{\text{T},\mu,:}\right) = \frac{\sum_{l=1}^{L} \frac{\phi_{c_{x,l}} \mu^2}{\left(\mu + \lambda_{\max,l}\right)^2}}{\sum_{l=1}^{L} \phi_{c_{x,l}}}. \tag{6.77}$$

Property 6.5 The fullband speech-distortion index of the tradeoff filter is an increasing function of the parameter μ.

Proof. It is straightforward to verify that

$$\frac{d\upsilon_{\text{sd}}\left(\mathbf{h}_{\text{T},\mu,:}\right)}{d\mu} \geq 0, \tag{6.78}$$

which ends the proof. □

It is clear that

$$0 \leq \upsilon_{\text{sd}}\left(\mathbf{h}_{\text{T},\mu,:}\right) \leq 1, \ \forall \mu \geq 0. \tag{6.79}$$

Therefore, as μ increases, the fullband output SNR increases at the price of more distortion to the desired signal.

Property 6.6 With the tradeoff filter, $\mathbf{h}_{\text{T},\mu,l}$, the fullband output SNR is always greater than or equal to the input SNR, i.e., $\text{oSNR}\left(\mathbf{h}_{\text{T},\mu,:}\right) \geq \text{iSNR}, \ \forall \mu \geq 0$.

Proof. We know that

$$\lambda_{\max,l} \geq \text{iSNR}_l, \tag{6.80}$$

which implies that

$$\sum_{l=1}^{L} \phi_{c_{v,l}} \frac{\text{iSNR}_l}{\lambda_{\max,l}} \leq \sum_{l=1}^{L} \phi_{c_{v,l}} \tag{6.81}$$

and hence,

$$\text{oSNR}\left(\mathbf{h}_{\text{T},0,:}\right) = \frac{\sum_{l=1}^{L} \phi_{c_{x,l}}}{\sum_{l=1}^{L} \phi_{c_{v,l}} \frac{\text{iSNR}_l}{\lambda_{\max,l}}} \geq \frac{\sum_{l=1}^{L} \phi_{c_{x,l}}}{\sum_{l=1}^{L} \phi_{c_{v,l}}} = \text{iSNR}. \tag{6.82}$$

But from Proposition 6.3, we have

$$\text{oSNR}\left(\mathbf{h}_{\text{T},\mu,:}\right) \geq \text{oSNR}\left(\mathbf{h}_{\text{T},0,:}\right), \ \forall \mu \geq 0, \tag{6.83}$$

as a result,

$$\text{oSNR}\left(\mathbf{h}_{\text{T},\mu,:}\right) \geq \text{iSNR}, \ \forall \mu \geq 0, \tag{6.84}$$

which completes the proof. □

CHAPTER 7

Optimal Filters in the KLE Domain with Model 3

This chapter is dedicated to the study of optimal filters with Model 3 where the interband correlation is taken into account. To simplify the presentation, we drop the subscript "3" from the FIR filter of length L and the filtering matrix of size $L \times L$ (see Chapter 4, Section 4.3), so that now $\mathbf{h}_{3,l}$ and $\mathbf{H}_3$ are written as $\mathbf{h}_l$ and $\mathbf{H}$, respectively.

7.1 PERFORMANCE MEASURES

In this section, we derive the most important performance measures based on Model 3 derived in Chapter 4, Section 4.3.

We define the subband output SNR as

$$\begin{aligned} \text{oSNR}(\mathbf{h}_l) &= \frac{\mathbf{h}_l^T \mathbf{\Phi}_{\mathbf{c}_x} \mathbf{h}_l}{\mathbf{h}_l^T \mathbf{\Phi}_{\mathbf{c}_v} \mathbf{h}_l} \\ &= \frac{(\mathbf{Q}\mathbf{h}_l)^T \mathbf{R}_{\mathbf{x}} (\mathbf{Q}\mathbf{h}_l)}{(\mathbf{Q}\mathbf{h}_l)^T \mathbf{R}_{\mathbf{v}} (\mathbf{Q}\mathbf{h}_l)}, \ l = 1, 2, \ldots, L. \end{aligned} \tag{7.1}$$

It is interesting to notice that, contrary to Model 1 and Model 2 where the subband output SNRs depend only on the energies of the desired and noise signals in the considered subband, the subband output SNR for Model 3 depends on the whole energies (from all subbands) of the desired and noise signals.

We easily find the definition of the fullband output SNR, which is

$$\begin{aligned} \text{oSNR}(\mathbf{h}_:) &= \text{oSNR}(\mathbf{H}) \\ &= \frac{\sum_{l=1}^{L} \mathbf{h}_l^T \mathbf{\Phi}_{\mathbf{c}_x} \mathbf{h}_l}{\sum_{l=1}^{L} \mathbf{h}_l^T \mathbf{\Phi}_{\mathbf{c}_v} \mathbf{h}_l} \\ &= \frac{\text{tr}\left(\mathbf{H}\mathbf{\Phi}_{\mathbf{c}_x}\mathbf{H}^T\right)}{\text{tr}\left(\mathbf{H}\mathbf{\Phi}_{\mathbf{c}_v}\mathbf{H}^T\right)}. \end{aligned} \tag{7.2}$$

We recall that

$$\mathbf{H} = \begin{bmatrix} \mathbf{h}_1^T \\ \mathbf{h}_2^T \\ \vdots \\ \mathbf{h}_L^T \end{bmatrix}$$

is a filtering matrix of size $L \times L$.

We always have

$$\text{oSNR}(\mathbf{H}) \leq \sum_{l=1}^{L} \text{oSNR}(\mathbf{h}_l). \tag{7.3}$$

The previous inequality shows that the fullband output SNR is upper bounded no matter how the filters $\mathbf{h}_l$, $l = 1, 2, \ldots, L$ are chosen.

The subband and fullband noise-reduction factors are

$$\begin{aligned} \xi_{\text{nr}}(\mathbf{h}_l) &= \frac{\phi_{c_{v,l}}}{\mathbf{h}_l^T \mathbf{\Phi}_{\mathbf{c}_v} \mathbf{h}_l} \\ &= \frac{\mathbf{q}_l^T \mathbf{R}_{\mathbf{v}} \mathbf{q}_l}{(\mathbf{Q}\mathbf{h}_l)^T \mathbf{R}_{\mathbf{v}} (\mathbf{Q}\mathbf{h}_l)}, \ l = 1, 2, \ldots, L, \end{aligned} \tag{7.4}$$

$$\begin{aligned} \xi_{\text{nr}}(\mathbf{H}) &= \frac{\sum_{l=1}^{L} \phi_{c_{v,l}}}{\sum_{l=1}^{L} \mathbf{h}_l^T \mathbf{\Phi}_{\mathbf{c}_v} \mathbf{h}_l} \\ &= \frac{\text{tr}(\mathbf{R}_{\mathbf{v}})}{\text{tr}\left[\left(\mathbf{H}\mathbf{Q}^T\right) \mathbf{R}_{\mathbf{v}} \left(\mathbf{H}\mathbf{Q}^T\right)^T\right]}. \end{aligned} \tag{7.5}$$

These factors should be lower bounded by 1 for optimal filters. We also have

$$\xi_{\text{nr}}(\mathbf{H}) \leq \sum_{l=1}^{L} \xi_{\text{nr}}(\mathbf{h}_l). \tag{7.6}$$

The distortion of the desired signal can be quantified with the subband speech-distortion index

$$\begin{aligned} \upsilon_{\text{sd}}(\mathbf{h}_l) &= \frac{E\left\{\left[\mathbf{h}_l^T \mathbf{c}_x(m) - c_{x,l}(m)\right]^2\right\}}{\phi_{c_{x,l}}} \\ &= \frac{(\mathbf{h}_l - \mathbf{i}_l)^T \mathbf{\Phi}_{\mathbf{c}_x} (\mathbf{h}_l - \mathbf{i}_l)}{\phi_{c_{x,l}}}, \ l = 1, 2, \ldots, L \end{aligned} \tag{7.7}$$

and the fullband speech-distortion index

$$\begin{aligned} \upsilon_{\mathrm{sd}}(\mathbf{H}) &= \frac{\sum_{l=1}^{L}\left[\left(\mathbf{h}_l - \mathbf{i}_l\right)^T \boldsymbol{\Phi}_{\mathbf{c}_x}\left(\mathbf{h}_l - \mathbf{i}_l\right)\right]}{\sum_{l=1}^{L}\phi_{c_{x,l}}} \\ &= \frac{\mathrm{tr}\left[\left(\mathbf{H} - \mathbf{I}\right)\boldsymbol{\Phi}_{\mathbf{c}_x}\left(\mathbf{H} - \mathbf{I}\right)^T\right]}{\mathrm{tr}\left(\boldsymbol{\Phi}_{\mathbf{c}_x}\right)}, \end{aligned} \tag{7.8}$$

where $\mathbf{i}_l$ is a vector of length L, corresponding to the lth column of the identity matrix $\mathbf{I}$ of size $L \times L$. The speech-distortion index is usually upper bounded by 1. We have

$$\upsilon_{\mathrm{sd}}(\mathbf{H}) \leq \sum_{l=1}^{L} \upsilon_{\mathrm{sd}}(\mathbf{h}_l). \tag{7.9}$$

We can also quantify signal distortion via the subband and fullband speech-reduction factors which are defined as

$$\begin{aligned} \xi_{\mathrm{sr}}(\mathbf{h}_l) &= \frac{\phi_{c_{x,l}}}{\mathbf{h}_l^T \boldsymbol{\Phi}_{\mathbf{c}_x} \mathbf{h}_l} \\ &= \frac{\mathbf{q}_l^T \mathbf{R}_{\mathbf{x}} \mathbf{q}_l}{\left(\mathbf{Q}\mathbf{h}_l\right)^T \mathbf{R}_{\mathbf{x}} \left(\mathbf{Q}\mathbf{h}_l\right)}, \; l = 1, 2, \ldots, L, \end{aligned} \tag{7.10}$$

$$\begin{aligned} \xi_{\mathrm{sr}}(\mathbf{H}) &= \frac{\sum_{l=1}^{L}\phi_{c_{x,l}}}{\sum_{l=1}^{L}\mathbf{h}_l^T \boldsymbol{\Phi}_{\mathbf{c}_x} \mathbf{h}_l} \\ &= \frac{\mathrm{tr}\left(\mathbf{R}_{\mathbf{x}}\right)}{\mathrm{tr}\left[\left(\mathbf{H}\mathbf{Q}^T\right)\mathbf{R}_{\mathbf{x}}\left(\mathbf{H}\mathbf{Q}^T\right)^T\right]}. \end{aligned} \tag{7.11}$$

The speech-reduction factor is supposed to have a lower bound of 1 for optimal filters. We also have

$$\xi_{\mathrm{sr}}(\mathbf{H}) \leq \sum_{l=1}^{L} \xi_{\mathrm{sr}}(\mathbf{h}_l). \tag{7.12}$$

We can verify that

$$\frac{\mathrm{oSNR}(\mathbf{h}_l)}{\mathrm{iSNR}_l} = \frac{\xi_{\mathrm{nr}}(\mathbf{h}_l)}{\xi_{\mathrm{sr}}(\mathbf{h}_l)}, \; l = 1, 2, \ldots, L, \tag{7.13}$$

$$\frac{\mathrm{oSNR}(\mathbf{H})}{\mathrm{iSNR}} = \frac{\xi_{\mathrm{nr}}(\mathbf{H})}{\xi_{\mathrm{sr}}(\mathbf{H})}. \tag{7.14}$$

7.2 MSE CRITERION

We define the error signal between the estimated and desired signals in the subband l as

$$\begin{aligned} e_l(m) &= c_{z_3,l}(m) - c_{x,l}(m) \\ &= \mathbf{h}_l^T \mathbf{c}_y(m) - c_{x,l}(m). \end{aligned} \tag{7.15}$$

This error signal can also be written as the sum of two uncorrelated error signals:

$$e_l(m) = e_{x,l}(m) + e_{v,l}(m), \tag{7.16}$$

where

$$\begin{aligned} e_{x,l}(m) &= \mathbf{h}_l^T \mathbf{c}_x(m) - c_{x,l}(m) \\ &= \left(\mathbf{h}_l - \mathbf{i}_l\right)^T \mathbf{c}_x(m) \end{aligned} \tag{7.17}$$

is the speech distortion due to the filter and

$$e_{v,l}(m) = \mathbf{h}_l^T \mathbf{c}_v(m) \tag{7.18}$$

represents the residual noise.

From the error signal defined in (7.15), we can now deduce the subband MSE criterion:

$$\begin{aligned} J\left(\mathbf{h}_l\right) &= E\left[e_l^2(m)\right] \\ &= \mathbf{h}_l^T \mathbf{\Phi}_{\mathbf{c}_y} \mathbf{h}_l - 2\mathbf{h}_l^T \mathbf{\Phi}_{\mathbf{c}_y\mathbf{c}_x} \mathbf{i}_l + \phi_{c_{x,l}}, \end{aligned} \tag{7.19}$$

where

$$\begin{aligned} \mathbf{\Phi}_{\mathbf{c}_y\mathbf{c}_x} &= E\left[\mathbf{c}_y(m)\mathbf{c}_x^T(m)\right] \\ &= E\left[\mathbf{c}_x(m)\mathbf{c}_x^T(m)\right] \\ &= \mathbf{\Phi}_{\mathbf{c}_x} \end{aligned}$$

is the cross-correlation matrix between the two signal vectors $\mathbf{c}_y(m)$ and $\mathbf{c}_x(m)$.

The subband MSE can be rewritten as

$$J\left(\mathbf{h}_l\right) = J_x\left(\mathbf{h}_l\right) + J_v\left(\mathbf{h}_l\right),$$

where

$$\begin{aligned} J_x\left(\mathbf{h}_l\right) &= E\left[e_{x,l}^2(m)\right] \\ &= \left(\mathbf{h}_l - \mathbf{i}_l\right)^T \mathbf{\Phi}_{\mathbf{c}_x} \left(\mathbf{h}_l - \mathbf{i}_l\right) \end{aligned} \tag{7.20}$$

and

$$\begin{aligned} J_v\left(\mathbf{h}_l\right) &= E\left[e_{v,l}^2(m)\right] \\ &= \mathbf{h}_l^T \mathbf{\Phi}_{\mathbf{c}_v} \mathbf{h}_l. \end{aligned} \tag{7.21}$$

For the particular filters $\mathbf{h}_l = \mathbf{i}_l, \ \forall l$, we get

$$J(\mathbf{i}_l) = \phi_{C_{v,l}}. \tag{7.22}$$

Using this particular case of the MSE, we define the subband normalized MSE (NMSE) as

$$\begin{aligned} \tilde{J}\left(\mathbf{h}_l\right) &= \frac{J\left(\mathbf{h}_l\right)}{J\left(\mathbf{i}_l\right)} \\ &= \mathrm{iSNR}_l \cdot \upsilon_{\mathrm{sd}}\left(\mathbf{h}_l\right) + \frac{1}{\xi_{\mathrm{nr}}\left(\mathbf{h}_l\right)}, \end{aligned} \tag{7.23}$$

where

$$\upsilon_{\mathrm{sd}}\left(\mathbf{h}_l\right) = \frac{J_x\left(\mathbf{h}_l\right)}{\phi_{C_{x,l}}}, \tag{7.24}$$

$$\xi_{\mathrm{nr}}\left(\mathbf{h}_l\right) = \frac{\phi_{C_{v,l}}}{J_v\left(\mathbf{h}_l\right)}. \tag{7.25}$$

The KLE-domain NMSE depends explicitly on the subband speech-distortion index and the subband noise-reduction factor.

We define the fullband MSE and fullband NMSE as

$$\begin{aligned} J(\mathbf{H}) &= \frac{1}{L}\sum_{l=1}^{L} J\left(\mathbf{h}_l\right) \\ &= \frac{1}{L}\sum_{l=1}^{L} J_x\left(\mathbf{h}_l\right) + \frac{1}{L}\sum_{l=1}^{L} J_v\left(\mathbf{h}_l\right) \\ &= J_x(\mathbf{H}) + J_v(\mathbf{H}) \end{aligned} \tag{7.26}$$

and

$$\begin{aligned} \tilde{J}(\mathbf{H}) &= L\frac{J(\mathbf{H})}{\sum_{l=1}^{L}\phi_{C_{v,l}}} \\ &= \mathrm{iSNR} \cdot \upsilon_{\mathrm{sd}}(\mathbf{H}) + \frac{1}{\xi_{\mathrm{nr}}(\mathbf{H})}, \end{aligned} \tag{7.27}$$

where

$$\upsilon_{\mathrm{sd}}(\mathbf{H}) = \frac{J_x(\mathbf{H})}{\sum_{l=1}^{L}\phi_{C_{x,l}}}, \tag{7.28}$$

$$\xi_{\mathrm{nr}}(\mathbf{H}) = \frac{\sum_{l=1}^{L}\phi_{C_{v,l}}}{J_v(\mathbf{H})}. \tag{7.29}$$

The fullband NMSE with the KLE depends also explicitly on the fullband speech-distortion index and the fullband noise-reduction factor.

It is straightforward to see that minimizing the subband MSE for each l is equivalent to minimizing the fullband MSE.

7.3 WIENER FILTER

If we differentiate the MSE criterion, $J(\mathbf{h}_l)$, with respect to $\mathbf{h}_l$ and equate the result to zero, we find the Wiener filter:

$$\begin{aligned} \mathbf{h}_{\mathrm{W},l} &= \mathbf{\Phi}_{\mathbf{c}_y}^{-1}\mathbf{\Phi}_{\mathbf{c}_x}\mathbf{i}_l \\ &= \left(\mathbf{I} - \mathbf{\Phi}_{\mathbf{c}_y}^{-1}\mathbf{\Phi}_{\mathbf{c}_v}\right)\mathbf{i}_l \\ &= \left(\mathbf{I} - \mathbf{\Lambda}^{-1}\mathbf{\Phi}_{\mathbf{c}_v}\right)\mathbf{i}_l. \end{aligned} \tag{7.30}$$

Combining all filters $\mathbf{h}_{\mathrm{W},l},\ l = 1, 2, \ldots, L$ in a matrix, we get

$$\begin{aligned} \mathbf{H}_{\mathrm{W}} &= \mathbf{\Phi}_{\mathbf{c}_x}\mathbf{\Phi}_{\mathbf{c}_y}^{-1} \\ &= \mathbf{I} - \mathbf{\Phi}_{\mathbf{c}_v}\mathbf{\Phi}_{\mathbf{c}_y}^{-1} \\ &= \mathbf{I} - \mathbf{\Phi}_{\mathbf{c}_v}\mathbf{\Lambda}^{-1}. \end{aligned} \tag{7.31}$$

Property 7.1 The Wiener filter derived with Model 3 [eq. (7.31)] is strictly equivalent to the classical time-domain Wiener filter derived in Chapter 3 [eq. (3.22)].

Proof. Indeed, from Chapter 4, Section 4.3, we know that the time-domain form of $\mathbf{H}_{\mathrm{W}}$ is

$$\mathbf{H}_{\mathrm{TD,W}} = \mathbf{Q}\mathbf{H}_{\mathrm{W}}\mathbf{Q}^T. \tag{7.32}$$

Substituting (7.31) into the previous expression, we find that

$$\begin{aligned} \mathbf{H}_{\mathrm{TD,W}} &= \mathbf{Q}\mathbf{\Phi}_{\mathbf{c}_x}\mathbf{\Lambda}^{-1}\mathbf{Q}^T \\ &= \left(\mathbf{Q}\mathbf{\Phi}_{\mathbf{c}_x}\mathbf{Q}^T\right)\left(\mathbf{Q}\mathbf{\Lambda}^{-1}\mathbf{Q}^T\right) \\ &= \mathbf{R}_{\mathbf{x}}\mathbf{R}_{\mathbf{y}}^{-1} \\ &= \mathbf{H}_{\mathrm{t,W}}, \end{aligned} \tag{7.33}$$

which completes the proof. □

It is interesting to see that the subband and fullband output SNRs are

$$\mathrm{oSNR}(\mathbf{h}_{\mathrm{W},l}) = \frac{\mathbf{q}_l^T\mathbf{R}_{\mathbf{x}}\mathbf{R}_{\mathbf{y}}^{-1}\mathbf{R}_{\mathbf{x}}\mathbf{R}_{\mathbf{y}}^{-1}\mathbf{R}_{\mathbf{x}}\mathbf{q}_l}{\mathbf{q}_l^T\mathbf{R}_{\mathbf{x}}\mathbf{R}_{\mathbf{y}}^{-1}\mathbf{R}_{\mathbf{v}}\mathbf{R}_{\mathbf{y}}^{-1}\mathbf{R}_{\mathbf{x}}\mathbf{q}_l},\ l = 1, 2, \ldots, L, \tag{7.34}$$

$$\begin{aligned} \mathrm{oSNR}(\mathbf{H}_{\mathrm{W}}) &= \mathrm{oSNR}\left(\mathbf{H}_{\mathrm{TD,W}}\right) \\ &= \mathrm{oSNR}\left(\mathbf{H}_{\mathrm{t,W}}\right) \\ &= \frac{\mathrm{tr}\left(\mathbf{R}_{\mathbf{x}}\mathbf{R}_{\mathbf{y}}^{-1}\mathbf{R}_{\mathbf{x}}\mathbf{R}_{\mathbf{y}}^{-1}\mathbf{R}_{\mathbf{x}}\right)}{\mathrm{tr}\left(\mathbf{R}_{\mathbf{x}}\mathbf{R}_{\mathbf{y}}^{-1}\mathbf{R}_{\mathbf{v}}\mathbf{R}_{\mathbf{y}}^{-1}\mathbf{R}_{\mathbf{x}}\right)}. \end{aligned} \tag{7.35}$$

From some results of this chapter and Chapter 3, we easily deduce that

$$\text{oSNR}(\mathbf{H}_\text{W}) \geq \text{iSNR} \tag{7.36}$$

and

$$\text{oSNR}(\mathbf{H}_\text{W}) \leq \sum_{l=1}^{L} \frac{\mathbf{q}_l^T \mathbf{R}_\mathbf{x} \mathbf{R}_\mathbf{y}^{-1} \mathbf{R}_\mathbf{x} \mathbf{R}_\mathbf{y}^{-1} \mathbf{R}_\mathbf{x} \mathbf{q}_l}{\mathbf{q}_l^T \mathbf{R}_\mathbf{x} \mathbf{R}_\mathbf{y}^{-1} \mathbf{R}_\mathbf{v} \mathbf{R}_\mathbf{y}^{-1} \mathbf{R}_\mathbf{x} \mathbf{q}_l}. \tag{7.37}$$

7.4 TRADEOFF FILTER

The basic principle of the tradeoff filter is to compromise between noise reduction and speech distortion. From the following optimization procedure

$$\min_{\mathbf{h}_l} J_x\left(\mathbf{h}_l\right) \quad \text{subject to} \quad J_v\left(\mathbf{h}_l\right) = \beta \phi_{c_{v,l}}, \tag{7.38}$$

where $0 < \beta < 1$ to insure that we get some noise reduction, we find that the optimal tradeoff filter is

$$\mathbf{h}_{\text{T},\mu,l} = \left(\mathbf{\Phi}_{\mathbf{c}_x} + \mu \mathbf{\Phi}_{\mathbf{c}_v}\right)^{-1} \mathbf{\Phi}_{\mathbf{c}_x} \mathbf{i}_l, \tag{7.39}$$

where $\mu \geq 0$ is a Lagrange multiplier satisfying $J_v\left(\mathbf{h}_{\text{T},\mu,l}\right) = \beta \phi_{c_{v,l}}$. Usually μ is chosen in an ad-hoc way, so that for

- $\mu = 1$, $\mathbf{h}_{\text{T},1,l} = \mathbf{h}_{\text{W},l}$, which is the Wiener filter;
- $\mu = 0$, $\mathbf{h}_{\text{T},0,l} = \mathbf{i}_l$, which is the identity filter (neither noise reduction nor speech distortion);
- $\mu > 1$, results in low residual noise at the expense of high speech distortion;
- $\mu < 1$, results in high residual noise and low speech distortion.

Combining all filters $\mathbf{h}_{\text{T},\mu,l}$, $l = 1, 2, \ldots, L$ in a matrix, we get

$$\mathbf{H}_{\text{T},\mu} = \mathbf{\Phi}_{\mathbf{c}_x} \left(\mathbf{\Phi}_{\mathbf{c}_x} + \mu \mathbf{\Phi}_{\mathbf{c}_v}\right)^{-1}. \tag{7.40}$$

Property 7.2 The tradeoff filter derived with Model 3 [eq. (7.40)] is strictly equivalent to the classical time-domain tradeoff filter derived in Chapter 3 [eq. (3.47)].

Proof. Indeed, from Chapter 4, Section 4.3, we know that the time-domain form of $\mathbf{H}_{\text{T},\mu}$ is

$$\mathbf{H}_{\text{TD,T},\mu} = \mathbf{Q}\mathbf{H}_{\text{T},\mu}\mathbf{Q}^T. \tag{7.41}$$

Substituting (7.40) into the previous expression, we find that

$$\begin{aligned} \mathbf{H}_{\mathrm{TD,T},\mu} &= \left(\mathbf{Q}\boldsymbol{\Phi}_{\mathbf{c}_x}\mathbf{Q}^T\right)\mathbf{Q}\left(\boldsymbol{\Phi}_{\mathbf{c}_x} + \mu\boldsymbol{\Phi}_{\mathbf{c}_v}\right)^{-1}\mathbf{Q}^T \\ &= \mathbf{R}_{\mathbf{x}}\left(\mathbf{R}_{\mathbf{x}} + \mu\mathbf{R}_{\mathbf{v}}\right)^{-1} \\ &= \mathbf{H}_{\mathrm{t,T},\mu}, \end{aligned} \tag{7.42}$$

which completes the proof. □

Obviously, we also have the following important results

$$\begin{aligned} \mathrm{oSNR}(\mathbf{H}_{\mathrm{T},\mu}) &= \mathrm{oSNR}\left(\mathbf{H}_{\mathrm{TD,T},\mu}\right) \\ &= \mathrm{oSNR}\left(\mathbf{H}_{\mathrm{t,T},\mu}\right) \end{aligned} \tag{7.43}$$

and

$$\mathrm{oSNR}(\mathbf{H}_{\mathrm{T},\mu}) \geq \mathrm{iSNR}, \ \forall \mu \geq 0. \tag{7.44}$$

7.5 MAXIMUM SNR FILTER

The maximum SNR filter is obtained by maximizing the subband output SNR defined in (7.1). Assume that the matrix $\boldsymbol{\Phi}_{\mathbf{c}_v}$ is full rank. In this case, the maximum SNR filter is the same in all subbands and is equal to the eigenvector, $\mathbf{h}_{\max}$, corresponding to the maximum eigenvalue, $\lambda_{\max}$, of the matrix $\boldsymbol{\Phi}_{\mathbf{c}_v}^{-1}\boldsymbol{\Phi}_{\mathbf{c}_x}$. As a result, the subband and fullband output SNRs are

$$\mathrm{oSNR}\left(\mathbf{h}_{\max}\right) = \lambda_{\max}, \ \forall l, \tag{7.45}$$

$$\mathrm{oSNR}\left(\mathbf{H}_{\max}\right) = \lambda_{\max}, \tag{7.46}$$

where

$$\mathbf{H}_{\max} = \begin{bmatrix} \beta_1 \mathbf{h}_{\max}^T \\ \beta_2 \mathbf{h}_{\max}^T \\ \vdots \\ \beta_L \mathbf{h}_{\max}^T \end{bmatrix} \tag{7.47}$$

and β_l, $l = 1, 2, \ldots, L$ are real numbers with at least one of them different from 0.

Property 7.3 The maximum SNR filter derived with Model 3 [eq. (7.47)] is strictly equivalent to the time-domain maximum SNR filter derived in Chapter 3 [eq. (3.74)].

Proof. Indeed, from Chapter 4, Section 4.3, we know that the time-domain form of $\mathbf{H}_{\max}$ is

$$\mathbf{H}_{\mathrm{TD},\max} = \mathbf{Q}\mathbf{H}_{\max}\mathbf{Q}^T. \tag{7.48}$$

It is clear that

$$\mathbf{R}_{\mathbf{v}}^{-1}\mathbf{R}_{\mathbf{x}}\mathbf{h}_{\mathrm{t,max}} = \lambda_{\mathrm{max}}\mathbf{h}_{\mathrm{t,max}}, \tag{7.49}$$

$$\mathbf{\Phi}_{\mathbf{c}_v}^{-1}\mathbf{\Phi}_{\mathbf{c}_x}\mathbf{h}_{\mathrm{max}} = \lambda_{\mathrm{max}}\mathbf{h}_{\mathrm{max}}, \tag{7.50}$$

where

$$\mathbf{h}_{\mathrm{max}} = \mathbf{Q}^T\mathbf{h}_{\mathrm{t,max}}. \tag{7.51}$$

Therefore, substituting (7.51) into (7.48), we find that

$$\mathbf{H}_{\mathrm{TD,max}} = \mathbf{Q}\mathbf{H}_{\mathrm{t,max}} = \mathbf{H}'_{\mathrm{t,max}}. \tag{7.52}$$

□

This is another interesting (and simpler) way to derive the maximum SNR filter in the time domain as compared to its direct derivation from the time-domain output SNR as explained in Chapter 3.

CHAPTER 8

Optimal Filters in the KLE Domain with Model 4

Model 4 is at the same time the most general model and the most complicated one since both the interband and interframe correlations are taken into account. To simplify the presentation, we drop the subscript "4" from the FIR filter of length ML and the filtering matrix of size $L \times ML$ (see Chapter 4, Section 4.4), so that now $\underline{\mathbf{h}}_{4,l}$ and $\underline{\mathbf{H}}_4$ are written as $\underline{\mathbf{h}}_l$ and $\underline{\mathbf{H}}$, respectively.

8.1 PERFORMANCE MEASURES

All performance measures are derived from expressions (4.37) and (4.42) of Section 4.4.

We define the subband output SNR as[1]

$$\begin{aligned} \text{oSNR}\left(\underline{\mathbf{h}}_l\right) &= \frac{\underline{\mathbf{h}}_l^T \mathbf{\Phi}_{\underline{\mathbf{c}}_{x_\text{d}}} \underline{\mathbf{h}}_l}{\underline{\mathbf{h}}_l^T \mathbf{\Phi}_\text{in} \underline{\mathbf{h}}_l} \\ &= \frac{\underline{\mathbf{h}}_l^T \underline{\mathbf{\Gamma}}_{\mathbf{c}_x} \mathbf{\Phi}_{\mathbf{c}_x} \underline{\mathbf{\Gamma}}_{\mathbf{c}_x}^T \underline{\mathbf{h}}_l}{\underline{\mathbf{h}}_l^T \mathbf{\Phi}_\text{in} \underline{\mathbf{h}}_l}, \ l = 1, 2, \ldots, L, \end{aligned} \tag{8.1}$$

where

$$\mathbf{\Phi}_\text{in} = \mathbf{\Phi}_{\underline{\mathbf{c}}_x''} + \mathbf{\Phi}_{\underline{\mathbf{c}}_v} \tag{8.2}$$

is the interference-plus-noise covariance matrix. Like in Model 3, this subband output SNR depends on the whole energies (from all subbands) of the desired, interference, and noise signals.

The fullband output SNR is then

$$\begin{aligned} \text{oSNR}\left(\underline{\mathbf{h}}_{:}\right) &= \text{oSNR}\left(\underline{\mathbf{H}}\right) \\ &= \frac{\sum_{l=1}^{L} \underline{\mathbf{h}}_l^T \mathbf{\Phi}_{\underline{\mathbf{c}}_{x_\text{d}}} \underline{\mathbf{h}}_l}{\sum_{l=1}^{L} \underline{\mathbf{h}}_l^T \mathbf{\Phi}_\text{in} \underline{\mathbf{h}}_l} \\ &= \frac{\text{tr}\left(\underline{\mathbf{H}} \mathbf{\Phi}_{\underline{\mathbf{c}}_{x_\text{d}}} \underline{\mathbf{H}}^T\right)}{\text{tr}\left(\underline{\mathbf{H}} \mathbf{\Phi}_\text{in} \underline{\mathbf{H}}^T\right)}. \end{aligned} \tag{8.3}$$

[1]In this study, we consider the interference as part of the noise in the definitions of the performance measures.

We recall that

$$\underline{\mathbf{H}} = \begin{bmatrix} \underline{\mathbf{h}}_1^T \\ \underline{\mathbf{h}}_2^T \\ \vdots \\ \underline{\mathbf{h}}_L^T \end{bmatrix}$$

is a filtering matrix of size $L \times ML$.

We always have

$$\text{oSNR}\left(\underline{\mathbf{H}}\right) \leq \sum_{l=1}^{L} \text{oSNR}\left(\underline{\mathbf{h}}_l\right). \tag{8.4}$$

The previous inequality shows that the fullband output SNR is upper bounded no matter how the filters $\underline{\mathbf{h}}_l,\ l = 1, 2, \ldots, L$ are taken.

The subband and fullband noise-reduction factors are

$$\xi_{\text{nr}}\left(\underline{\mathbf{h}}_l\right) = \frac{\phi_{c_{v,l}}}{\underline{\mathbf{h}}_l^T \mathbf{\Phi}_{\text{in}} \underline{\mathbf{h}}_l},\ l = 1, 2, \ldots, L, \tag{8.5}$$

$$\begin{aligned} \xi_{\text{nr}}\left(\underline{\mathbf{H}}\right) &= \frac{\sum_{l=1}^{L} \phi_{c_{v,l}}}{\sum_{l=1}^{L} \underline{\mathbf{h}}_l^T \mathbf{\Phi}_{\text{in}} \underline{\mathbf{h}}_l} \\ &= \frac{\text{tr}\left(\mathbf{\Phi}_{\mathbf{c}_v}\right)}{\text{tr}\left(\underline{\mathbf{H}} \mathbf{\Phi}_{\text{in}} \underline{\mathbf{H}}^T\right)}. \end{aligned} \tag{8.6}$$

These factors should be lower bounded by 1 for optimal filters. We also have

$$\xi_{\text{nr}}\left(\underline{\mathbf{H}}\right) \leq \sum_{l=1}^{L} \xi_{\text{nr}}\left(\underline{\mathbf{h}}_l\right). \tag{8.7}$$

The distortion of the desired signal can be quantified with the subband speech-distortion index

$$\begin{aligned} \upsilon_{\text{sd}}\left(\underline{\mathbf{h}}_l\right) &= \frac{E\left\{\left[\underline{\mathbf{h}}_l^T \mathbf{\Gamma}_{\mathbf{c}_x} \mathbf{c}_x(m) - c_{x,l}(m)\right]^2\right\}}{\phi_{c_{x,l}}} \\ &= \frac{\left(\mathbf{\Gamma}_{\mathbf{c}_x}^T \underline{\mathbf{h}}_l - \mathbf{i}_l\right)^T \mathbf{\Phi}_{\mathbf{c}_x} \left(\mathbf{\Gamma}_{\mathbf{c}_x}^T \underline{\mathbf{h}}_l - \mathbf{i}_l\right)}{\phi_{c_{x,l}}},\ l = 1, 2, \ldots, L \end{aligned} \tag{8.8}$$

and the fullband speech-distortion index

$$\upsilon_{\text{sd}}\left(\underline{\mathbf{H}}\right) = \frac{\sum_{l=1}^{L}\left(\underline{\boldsymbol{\Gamma}}_{\mathbf{c}_x}^T\underline{\mathbf{h}}_l - \mathbf{i}_l\right)^T \boldsymbol{\Phi}_{\mathbf{c}_x}\left(\underline{\boldsymbol{\Gamma}}_{\mathbf{c}_x}^T\underline{\mathbf{h}}_l - \mathbf{i}_l\right)}{\sum_{l=1}^{L}\phi_{c_{x,l}}} = \frac{\text{tr}\left[\left(\underline{\mathbf{H}}\underline{\boldsymbol{\Gamma}}_{\mathbf{c}_x} - \mathbf{I}\right)\boldsymbol{\Phi}_{\mathbf{c}_x}\left(\underline{\mathbf{H}}\underline{\boldsymbol{\Gamma}}_{\mathbf{c}_x} - \mathbf{I}\right)^T\right]}{\text{tr}\left(\boldsymbol{\Phi}_{\mathbf{c}_x}\right)}, \tag{8.9}$$

where $\mathbf{i}_l$ is a vector of length L, corresponding to the lth column of the identity matrix $\mathbf{I}$ of size $L \times L$. The speech-distortion index is usually upper bounded by 1. We have

$$\upsilon_{\text{sd}}\left(\underline{\mathbf{H}}\right) \leq \sum_{l=1}^{L}\upsilon_{\text{sd}}\left(\underline{\mathbf{h}}_l\right). \tag{8.10}$$

We can also quantify signal distortion via the subband and fullband speech-reduction factors which are defined as

$$\xi_{\text{sr}}\left(\underline{\mathbf{h}}_l\right) = \frac{\phi_{c_{x,l}}}{\underline{\mathbf{h}}_l^T\underline{\boldsymbol{\Gamma}}_{\mathbf{c}_x}\boldsymbol{\Phi}_{\mathbf{c}_x}\underline{\boldsymbol{\Gamma}}_{\mathbf{c}_x}^T\underline{\mathbf{h}}_l},\ l = 1, 2, \ldots, L, \tag{8.11}$$

$$\xi_{\text{sr}}\left(\underline{\mathbf{H}}\right) = \frac{\sum_{l=1}^{L}\phi_{c_{x,l}}}{\sum_{l=1}^{L}\underline{\mathbf{h}}_l^T\underline{\boldsymbol{\Gamma}}_{\mathbf{c}_x}\boldsymbol{\Phi}_{\mathbf{c}_x}\underline{\boldsymbol{\Gamma}}_{\mathbf{c}_x}^T\underline{\mathbf{h}}_l} = \frac{\text{tr}\left(\boldsymbol{\Phi}_{\mathbf{c}_x}\right)}{\text{tr}\left(\underline{\mathbf{H}}\underline{\boldsymbol{\Gamma}}_{\mathbf{c}_x}\boldsymbol{\Phi}_{\mathbf{c}_x}\underline{\boldsymbol{\Gamma}}_{\mathbf{c}_x}^T\underline{\mathbf{H}}^T\right)}. \tag{8.12}$$

The speech-reduction factor is supposed to have a lower bound of 1 for optimal filters. We also have

$$\xi_{\text{sr}}\left(\underline{\mathbf{H}}\right) \leq \sum_{l=1}^{L}\xi_{\text{sr}}\left(\underline{\mathbf{h}}_l\right). \tag{8.13}$$

An important observation from (8.9) or (8.12) is that the design of a noise reduction algorithm with Model 4 that does not distort the desired signal requires the constraint

$$\underline{\mathbf{H}}\underline{\boldsymbol{\Gamma}}_{\mathbf{c}_x} = \mathbf{I}. \tag{8.14}$$

We can verify that

$$\frac{\text{oSNR}\left(\underline{\mathbf{h}}_l\right)}{\text{iSNR}_l} = \frac{\xi_{\text{nr}}\left(\underline{\mathbf{h}}_l\right)}{\xi_{\text{sr}}\left(\underline{\mathbf{h}}_l\right)},\ l = 1, 2, \ldots, L, \tag{8.15}$$

$$\frac{\text{oSNR}\left(\underline{\mathbf{H}}\right)}{\text{iSNR}} = \frac{\xi_{\text{nr}}\left(\underline{\mathbf{H}}\right)}{\xi_{\text{sr}}\left(\underline{\mathbf{H}}\right)}. \tag{8.16}$$

8.2 MSE CRITERION

The error signal between the estimated and desired signals in the subband l is defined as

$$\begin{aligned} e_l(m) &= c_{z_4,l}(m) - c_{x,l}(m) \\ &= \underline{\mathbf{h}}_l^T \underline{\mathbf{c}}_y(m) - c_{x,l}(m). \end{aligned} \tag{8.17}$$

The previous error can be decomposed as follows:

$$e_l(m) = e_{x,l}(m) + e_{\text{in},l}(m), \tag{8.18}$$

where

$$\begin{aligned} e_{x,l}(m) &= \underline{\mathbf{h}}_l^T \underline{\mathbf{\Gamma}}_{\mathbf{c}_x} \mathbf{c}_x(m) - c_{x,l}(m) \\ &= \left(\underline{\mathbf{\Gamma}}_{\mathbf{c}_x}^T \underline{\mathbf{h}}_l - \mathbf{i}_l \right)^T \mathbf{c}_x(m) \end{aligned} \tag{8.19}$$

is the speech distortion due to the filter and

$$e_{\text{in},l}(m) = \underline{\mathbf{h}}_l^T \underline{\mathbf{c}}_x''(m) + \underline{\mathbf{h}}_l^T \underline{\mathbf{c}}_v(m) \tag{8.20}$$

represents the residual interference-plus-noise.

The subband MSE criterion for Model 4 is then

$$\begin{aligned} J\left(\underline{\mathbf{h}}_l\right) &= E\left[e_l^2(m)\right] \\ &= \underline{\mathbf{h}}_l^T \mathbf{\Phi}_{\underline{\mathbf{c}}_y} \underline{\mathbf{h}}_l - 2\underline{\mathbf{h}}_l^T \mathbf{\Phi}_{\underline{\mathbf{c}}_y \underline{\mathbf{c}}_x} \underline{\mathbf{i}}_l + \phi_{c_{x,l}}, \end{aligned} \tag{8.21}$$

where

$$\mathbf{\Phi}_{\underline{\mathbf{c}}_y} = E\left[\underline{\mathbf{c}}_y(m) \underline{\mathbf{c}}_y^T(m)\right]$$

is the correlation matrix of the signal $\underline{\mathbf{c}}_y(m)$,

$$\begin{aligned} \mathbf{\Phi}_{\underline{\mathbf{c}}_y \underline{\mathbf{c}}_x} &= E\left[\underline{\mathbf{c}}_y(m) \underline{\mathbf{c}}_x^T(m)\right] \\ &= E\left[\underline{\mathbf{c}}_x(m) \underline{\mathbf{c}}_x^T(m)\right] \\ &= \mathbf{\Phi}_{\underline{\mathbf{c}}_x} \end{aligned}$$

is the cross-correlation matrix between the two signal vectors $\underline{\mathbf{c}}_y(m)$ and $\underline{\mathbf{c}}_x(m)$, and $\underline{\mathbf{i}}_l$ is a vector of length ML for which its lth component is equal to 1 and all its other components are equal to 0.

Expression (8.21) can be rewritten as

$$J\left(\underline{\mathbf{h}}_l\right) = J_x\left(\underline{\mathbf{h}}_l\right) + J_{\text{in}}\left(\underline{\mathbf{h}}_l\right),$$

where

$$\begin{aligned} J_x\left(\underline{\mathbf{h}}_l\right) &= E\left[e_{x,l}^2(m)\right] \\ &= \left(\underline{\mathbf{\Gamma}}_{\mathbf{c}_x}^T \underline{\mathbf{h}}_l - \underline{\mathbf{i}}_l\right)^T \mathbf{\Phi}_{\mathbf{c}_x} \left(\underline{\mathbf{\Gamma}}_{\mathbf{c}_x}^T \underline{\mathbf{h}}_l - \underline{\mathbf{i}}_l\right) \end{aligned} \tag{8.22}$$

and

$$\begin{aligned} J_{\text{in}}\left(\underline{\mathbf{h}}_l\right) &= E\left[e_{\text{in},l}^2(m)\right] \\ &= \underline{\mathbf{h}}_l^T \mathbf{\Phi}_{\text{in}} \underline{\mathbf{h}}_l. \end{aligned} \tag{8.23}$$

For the particular filters $\underline{\mathbf{h}}_l = \underline{\mathbf{i}}_l, \ \forall l$, we get

$$J\left(\underline{\mathbf{i}}_l\right) = \phi_{c_{v,l}}. \tag{8.24}$$

Using this particular case of the MSE, we define the subband normalized MSE (NMSE) as

$$\begin{aligned} \tilde{J}\left(\underline{\mathbf{h}}_l\right) &= \frac{J\left(\underline{\mathbf{h}}_l\right)}{J\left(\underline{\mathbf{i}}_l\right)} \\ &= \text{iSNR}_l \cdot \upsilon_{\text{sd}}\left(\underline{\mathbf{h}}_l\right) + \frac{1}{\xi_{\text{nr}}\left(\underline{\mathbf{h}}_l\right)}, \end{aligned} \tag{8.25}$$

where

$$\upsilon_{\text{sd}}\left(\underline{\mathbf{h}}_l\right) = \frac{J_x\left(\underline{\mathbf{h}}_l\right)}{\phi_{c_{x,l}}}, \tag{8.26}$$

$$\xi_{\text{nr}}\left(\underline{\mathbf{h}}_l\right) = \frac{\phi_{c_{v,l}}}{J_{\text{in}}\left(\underline{\mathbf{h}}_l\right)}. \tag{8.27}$$

The KLE-domain NMSE depends explicitly on the subband speech-distortion index and the subband noise-reduction factor.

We define the fullband MSE and fullband NMSE as

$$\begin{aligned} J\left(\underline{\mathbf{H}}\right) &= \frac{1}{L}\sum_{l=1}^{L} J\left(\underline{\mathbf{h}}_l\right) \\ &= \frac{1}{L}\sum_{l=1}^{L} J_x\left(\underline{\mathbf{h}}_l\right) + \frac{1}{L}\sum_{l=1}^{L} J_{\text{in}}\left(\underline{\mathbf{h}}_l\right) \\ &= J_x\left(\underline{\mathbf{H}}\right) + J_{\text{in}}\left(\underline{\mathbf{H}}\right) \end{aligned} \tag{8.28}$$

and

$$\begin{aligned} \tilde{J}\left(\underline{\mathbf{H}}\right) &= L\frac{J\left(\underline{\mathbf{H}}\right)}{\sum_{l=1}^{L}\phi_{c_{v,l}}} \\ &= \text{iSNR} \cdot \upsilon_{\text{sd}}\left(\underline{\mathbf{H}}\right) + \frac{1}{\xi_{\text{nr}}\left(\underline{\mathbf{H}}\right)}, \end{aligned} \tag{8.29}$$

where

$$\upsilon_{\rm sd}\left(\underline{\mathbf{H}}\right) = \frac{J_x\left(\underline{\mathbf{H}}\right)}{\sum_{l=1}^{L}\phi_{c_{x,l}}}, \tag{8.30}$$

$$\xi_{\rm nr}\left(\underline{\mathbf{H}}\right) = \frac{\sum_{l=1}^{L}\phi_{c_{v,l}}}{J_{\rm in}\left(\underline{\mathbf{H}}\right)}. \tag{8.31}$$

The fullband NMSE with the KLE depends also explicitly on the fullband speech-distortion index and the fullband noise-reduction factor.

It is straightforward to see that minimizing the subband MSE for each l is equivalent to minimizing the fullband MSE.

8.3 WIENER FILTER

From the MSE criterion given in (8.21), we easily derive the Wiener filter, which is

$$\begin{aligned} \underline{\mathbf{h}}_{\mathrm{W},l} &= \mathbf{\Phi}_{\underline{\mathbf{c}}_y}^{-1}\mathbf{\Phi}_{\underline{\mathbf{c}}_x}\underline{\mathbf{i}}_l \\ &= \left(\mathbf{I}_{ML} - \mathbf{\Phi}_{\underline{\mathbf{c}}_y}^{-1}\mathbf{\Phi}_{\underline{\mathbf{c}}_v}\right)\underline{\mathbf{i}}_l, \end{aligned} \tag{8.32}$$

where $\mathbf{I}_{ML}$ is the identity matrix of size $ML \times ML$. Combining all filters $\underline{\mathbf{h}}_{\mathrm{W},l}$, $l = 1, 2, \ldots, L$ in a matrix, we get

$$\begin{aligned} \underline{\mathbf{H}}_{\mathrm{W}} &= \underline{\mathbf{I}}\mathbf{\Phi}_{\underline{\mathbf{c}}_x}\mathbf{\Phi}_{\underline{\mathbf{c}}_y}^{-1} \\ &= \underline{\mathbf{I}} - \underline{\mathbf{I}}\mathbf{\Phi}_{\underline{\mathbf{c}}_v}\mathbf{\Phi}_{\underline{\mathbf{c}}_y}^{-1}, \end{aligned} \tag{8.33}$$

where

$$\underline{\mathbf{I}} = \left[\begin{array}{cc} \mathbf{I} & \mathbf{0}_{L\times(ML-L)} \end{array}\right].$$

Lemma 8.1 *We can rewrite the Wiener filter as*

$$\underline{\mathbf{H}}_{\mathrm{W}} = \left(\mathbf{\Phi}_{\mathbf{c}_x}^{-1} + \underline{\mathbf{\Gamma}}_{\mathbf{c}_x}^T\mathbf{\Phi}_{\rm in}^{-1}\underline{\mathbf{\Gamma}}_{\mathbf{c}_x}\right)^{-1}\underline{\mathbf{\Gamma}}_{\mathbf{c}_x}^T\mathbf{\Phi}_{\rm in}^{-1}. \tag{8.34}$$

Proof. This expression is easy to show by using the Woodbury's identity in the following decomposition

$$\mathbf{\Phi}_{\underline{\mathbf{c}}_y} = \underline{\mathbf{\Gamma}}_{\mathbf{c}_x}\mathbf{\Phi}_{\mathbf{c}_x}\underline{\mathbf{\Gamma}}_{\mathbf{c}_x}^T + \mathbf{\Phi}_{\rm in} \tag{8.35}$$

and replacing it in (8.33). □

The form of the Wiener filter presented in (8.34) is interesting because it shows an obvious link with some other optimal filters as it will be verified later.

The Wiener filter with Model 4 has also several interesting properties. For example, it can be shown that the fullband output SNR is always greater than or equal to the input SNR, i.e., $\text{oSNR}\left(\underline{\mathbf{H}}_{\text{W}}\right) \geq \text{iSNR}$.

8.4 TRADEOFF FILTER

The tradeoff filter is an elegant way to compromise between noise reduction and speech distortion. One natural approach for its derivation is as follows:

$$\min_{\underline{\mathbf{h}}_l} J_x\left(\underline{\mathbf{h}}_l\right) \quad \text{subject to} \quad J_{\text{in}}\left(\underline{\mathbf{h}}_l\right) = \beta \phi_{c_{v,l}}, \tag{8.36}$$

where $0 < \beta < 1$ to insure that we get some noise reduction. From (8.36), we find that the optimal tradeoff filter is

$$\underline{\mathbf{h}}_{\text{T},\mu,l} = \left(\mathbf{\Phi}_{\underline{\mathbf{c}}_x} + \mu \mathbf{\Phi}_{\text{in}}\right)^{-1} \mathbf{\Phi}_{\underline{\mathbf{c}}_x} \underline{\mathbf{i}}_l, \tag{8.37}$$

where $\mu > 0$ is a Lagrange multiplier satisfying $J_{\text{in}}\left(\underline{\mathbf{h}}_{\text{T},\mu,l}\right) = \beta \phi_{c_{v,l}}$. Taking $\mu = 1$, we obviously find the Wiener filter.

Combining all filters $\underline{\mathbf{h}}_{\text{T},\mu,l}$, $l = 1, 2, \ldots, L$ in a matrix, we get

$$\underline{\mathbf{H}}_{\text{T},\mu} = \underline{\mathbf{I}} \mathbf{\Phi}_{\underline{\mathbf{c}}_x} \left(\mathbf{\Phi}_{\underline{\mathbf{c}}_x} + \mu \mathbf{\Phi}_{\text{in}}\right)^{-1}, \tag{8.38}$$

which can be rewritten, thanks to the Woodbury's identity, as

$$\underline{\mathbf{H}}_{\text{T},\mu} = \left(\mu \mathbf{\Phi}_{\mathbf{c}_x}^{-1} + \underline{\mathbf{\Gamma}}_{\mathbf{c}_x}^T \mathbf{\Phi}_{\text{in}}^{-1} \underline{\mathbf{\Gamma}}_{\mathbf{c}_x}\right)^{-1} \underline{\mathbf{\Gamma}}_{\mathbf{c}_x}^T \mathbf{\Phi}_{\text{in}}^{-1}. \tag{8.39}$$

We can show here as well that

$$\text{oSNR}\left(\underline{\mathbf{H}}_{\text{T},\mu}\right) \geq \text{iSNR}, \ \forall \mu > 0. \tag{8.40}$$

8.5 MVDR FILTER

The minimum variance distortionless response (MVDR) filter is found by

$$\min_{\underline{\mathbf{H}}} \text{tr}\left(\underline{\mathbf{H}} \mathbf{\Phi}_{\text{in}} \underline{\mathbf{H}}^T\right) \quad \text{subject to} \quad \underline{\mathbf{H}} \underline{\mathbf{\Gamma}}_{\mathbf{c}_x} = \mathbf{I}. \tag{8.41}$$

Therefore, the optimal solution is

$$\underline{\mathbf{H}}_{\text{MVDR}} = \left(\underline{\mathbf{\Gamma}}_{\mathbf{c}_x}^T \mathbf{\Phi}_{\text{in}}^{-1} \underline{\mathbf{\Gamma}}_{\mathbf{c}_x}\right)^{-1} \underline{\mathbf{\Gamma}}_{\mathbf{c}_x}^T \mathbf{\Phi}_{\text{in}}^{-1}. \tag{8.42}$$

Lemma 8.2 *We can rewrite the MVDR filter as*

$$\underline{\mathbf{H}}_{\text{MVDR}} = \left(\underline{\boldsymbol{\Gamma}}_{\mathbf{c}_x}^T \boldsymbol{\Phi}_{\underline{\mathbf{c}}_y}^{-1} \underline{\boldsymbol{\Gamma}}_{\mathbf{c}_x}\right)^{-1} \underline{\boldsymbol{\Gamma}}_{\mathbf{c}_x}^T \boldsymbol{\Phi}_{\underline{\mathbf{c}}_y}^{-1}. \tag{8.43}$$

Proof. This expression is easy to show by using the Woodbury's identity in $\boldsymbol{\Phi}_{\underline{\mathbf{c}}_y}^{-1}$. □

From (8.43), it is easy to deduce the relationship between the MVDR and Wiener filters:

$$\underline{\mathbf{H}}_{\text{MVDR}} = \left(\underline{\mathbf{H}}_{\text{W}} \underline{\boldsymbol{\Gamma}}_{\mathbf{c}_x}\right)^{-1} \underline{\mathbf{H}}_{\text{W}}. \tag{8.44}$$

It can be shown that

$$\text{oSNR}\left(\underline{\mathbf{H}}_{\text{MVDR}}\right) \geq \text{iSNR}. \tag{8.45}$$

8.6 MAXIMUM SNR FILTER

The maximum SNR filter is obtained by maximizing the subband output SNR defined in (8.1). It is assumed that the matrix $\boldsymbol{\Phi}_{\text{in}}$ is full rank. In this case, the maximum SNR filter is the same in all subbands and is equal to the eigenvector, $\underline{\mathbf{h}}_{\max}$, corresponding to the maximum eigenvalue, $\underline{\lambda}_{\max}$, of the matrix $\boldsymbol{\Phi}_{\text{in}}^{-1} \boldsymbol{\Phi}_{\underline{\mathbf{c}}_{x_{\text{d}}}}$. As a result, the subband and fullband output SNRs are

$$\text{oSNR}\left(\underline{\mathbf{h}}_{\max}\right) = \underline{\lambda}_{\max}, \ \forall l, \tag{8.46}$$

$$\text{oSNR}\left(\underline{\mathbf{H}}_{\max}\right) = \underline{\lambda}_{\max}, \tag{8.47}$$

where

$$\underline{\mathbf{H}}_{\max} = \begin{bmatrix} \beta_1 \underline{\mathbf{h}}_{\max}^T \\ \beta_2 \underline{\mathbf{h}}_{\max}^T \\ \vdots \\ \beta_L \underline{\mathbf{h}}_{\max}^T \end{bmatrix} \tag{8.48}$$

and $\beta_l,\ l = 1, 2, \ldots, L$ are real numbers with at least one of them different from 0.

It can be observed that for $\mu \geq 1$,

$$\text{iSNR} \leq \text{oSNR}\left(\underline{\mathbf{H}}_{\text{MVDR}}\right) \leq \text{oSNR}\left(\underline{\mathbf{H}}_{\text{W}}\right) \leq \text{oSNR}\left(\underline{\mathbf{H}}_{\text{T},\mu}\right) \leq \text{oSNR}\left(\underline{\mathbf{H}}_{\max}\right) = \underline{\lambda}_{\max} \tag{8.49}$$

and for $\mu \leq 1$,

$$\text{iSNR} \leq \text{oSNR}\left(\underline{\mathbf{H}}_{\text{T},\mu}\right) \leq \text{oSNR}\left(\underline{\mathbf{H}}_{\text{MVDR}}\right) \leq \text{oSNR}\left(\underline{\mathbf{H}}_{\text{W}}\right) \leq \text{oSNR}\left(\underline{\mathbf{H}}_{\max}\right) = \underline{\lambda}_{\max}. \tag{8.50}$$

CHAPTER 9

Experimental Study

By dividing the general speech enhancement problem in the KLE domain into four categories, depending on whether the interframe and interband information is accounted for, we have derived a number of optimal noise reduction filters in Chapters 5 to 8. For each category of filters, we have analyzed their performance through theoretical evaluation of either the subband or the fullband output SNRs, noise-reduction factors, and speech-distortion indices. We have also discussed their connection to the time-domain filters. In this chapter, we study some of those key noise reduction filters through experiments and highlight the merits and limitations inherent in each optimal filter.

9.1 EXPERIMENTAL CONDITIONS

The clean speech used in all the experiments was recorded from a female talker in a quiet office room. It was sampled at 8 kHz. The overall length of the signal is 2 minutes. The first 10 seconds of this clean speech signal and the corresponding spectrogram are visualized in Fig. 9.1.

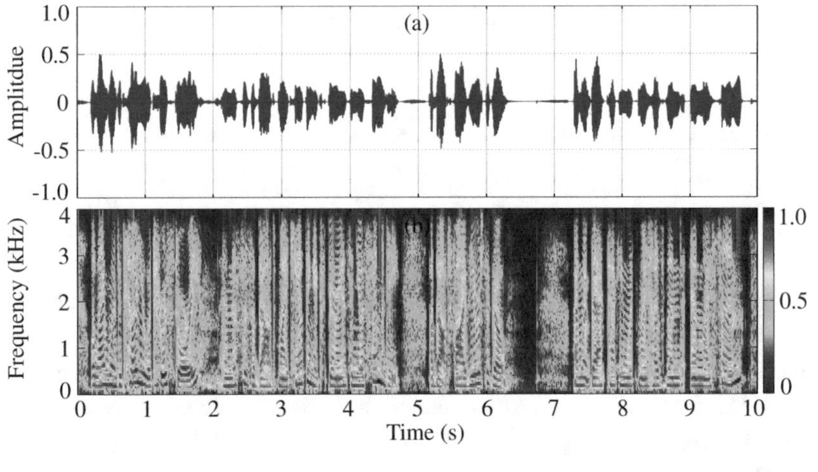

Figure 9.1: The clean speech $x(k)$ used in the experiments: (a) the first 10-second waveform and (b) the first 10-second spectrogram.

The noisy speech is obtained by adding noise to the clean speech where the noise signal is properly scaled to control the input SNR level. As we pointed out in the introduction, noise is a general term encompassing a broad range of unwanted signals. They can be either white (their spectral density is the same across all the frequency bands within the signal bandwidth) or colored (their power is not the same at different frequency bands); they can also be either stationary (their statistics stay the same over time) or nonstationary (their statistics are time-varying). Because of this, it is very difficult to evaluate a noise reduction filter and fairly compare different filters as the experimental results obtained in one noise condition may not necessarily be consistent with the ones obtained from another noise condition. Therefore, it is important to assess a noise reduction filter in many different conditions before we choose to implement it into a practical system. In this chapter, we choose three types of noise that we consider as very representative of real applications: a computer generated stationary white Gaussian random process, a car noise signal (quasi-stationary but colored), and a babble noise signal (nonstationary and colored).

The car noise is recorded in a Volvo Sedan running at 55 miles/hour on a highway with all its windows closed. The first 10 seconds of this noise and its spectrogram are shown in Fig. 9.2. It is seen from the spectrogram that most of the car noise energy concentrates in low frequencies so this noise is colored. Also plotted in Fig. 9.2 are the autocorrelation coefficients of this noise (the first column of $\mathbf{R}_\mathbf{v}$) computed using a long-time average. Clearly, there is a strong correlation between adjacent noise samples. This, again, illustrates that this car noise is colored.

The babble noise is recorded in a New York Stock Exchange (NYSE) room, so we shall call it the NYSE noise from now on. This noise consists of many sounds from various sources such as electrical fans, computer fans, telephone rings, and even some background speech. The first 10-second waveform and spectrogram and the first 20 autocorrelation coefficients (computed using a long-time average) of this noise are plotted in Fig. 9.3. One can easily see that the NYSE noise is nonstationary and colored.

9.2 ESTIMATION OF THE CORRELATION MATRICES AND VECTORS

The implementation of both the time-domain and KLE-domain noise reduction filters requires the estimation of the correlation matrices $\mathbf{R}_\mathbf{y}$, $\mathbf{R}_\mathbf{x}$, and $\mathbf{R}_\mathbf{v}$. Since the noisy signal $y(k)$ is accessible, the correlation matrix $\mathbf{R}_\mathbf{y}$ can be computed by approximating the mathematical expectation in its definition with a sample average. However, a noise estimator or a voice activity detector (VAD) is needed in practice to compute the other two matrices. While they are very important, the noise estimation and VAD issues will be left to the reader's investigation. Instead, we directly compute the noise correlation matrix from the noise signal. Specifically, in the following experiments, estimates of the matrices $\mathbf{R}_\mathbf{y}$ and $\mathbf{R}_\mathbf{v}$ are obtained using either a short-time or long-time average. In the short-time average case, at each time instant k, a segment of the noisy and noise signals that consists of a number of the most recent samples are used to compute the corresponding correlation matrices. The length of the segment (or window), denoted by N, may vary depending on the experimental setup, which will be specified in each experiment. In the long-time average, all the signal samples will be used to compute the correlation matrices. Once an estimate of the matrices $\mathbf{R}_\mathbf{y}$ and $\mathbf{R}_\mathbf{v}$ are achieved, the estimate of $\mathbf{R}_\mathbf{x}$ is obtained by subtracting the estimate of $\mathbf{R}_\mathbf{v}$ from that of $\mathbf{R}_\mathbf{y}$.

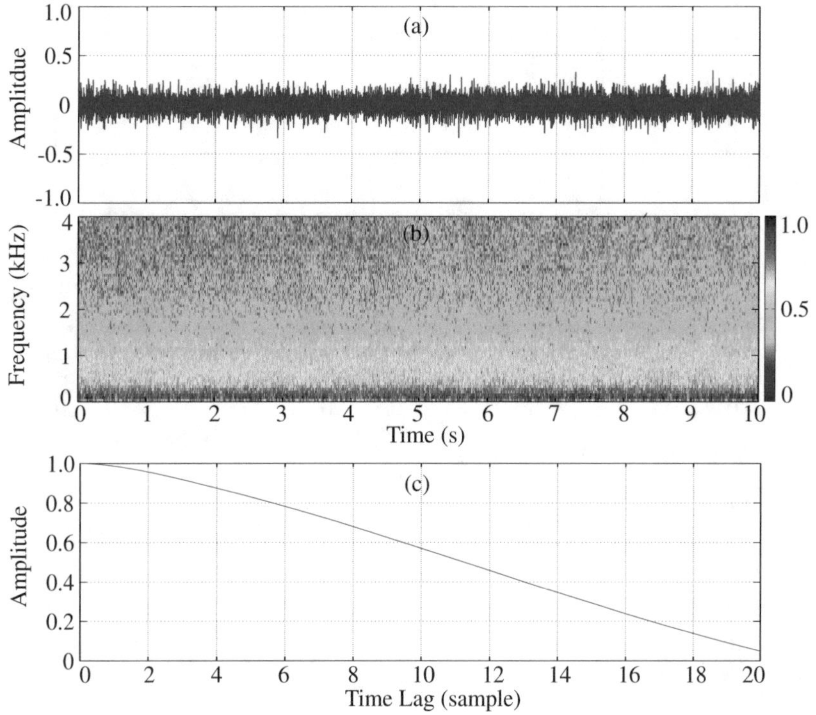

Figure 9.2: The car noise used in the experiments: (a) the first 10-second waveform, (b) the first 10-second spectrogram, and (c) the first 20 autocorrelation coefficients.

In the KLE domain, we also need to estimate the variances, correlation matrices, and correlation vectors of the subband signals $c_{y,l}$, $c_{x,l}$, and $c_{v,l}$. All the parameters of the noisy signal are computed from $c_{y,l}$ using a recursive method, and those parameters of the noise signal are directly computed from $c_{v,l}$ without using any VAD. The parameters associated with the clean speech are estimated by subtracting the corresponding noise parameters from those of the noisy signal.

9.3 PERFORMANCE MEASURES

For ease of comparison, we evaluate both the time- and KLE-domain filters using the fullband output SNR, noise-reduction factor, and speech-distortion index as the performance measures. These measures are computed according to their definitions given, respectively, in (3.7), (3.8), and (3.9) by replacing the expectation by a long-time average. Note that the KLE-domain filters are designed on a subband basis. To compute the fullband performance measures for these filters, we need to have the time-domain filtered speech, residual noise, and interference signal (if any), which

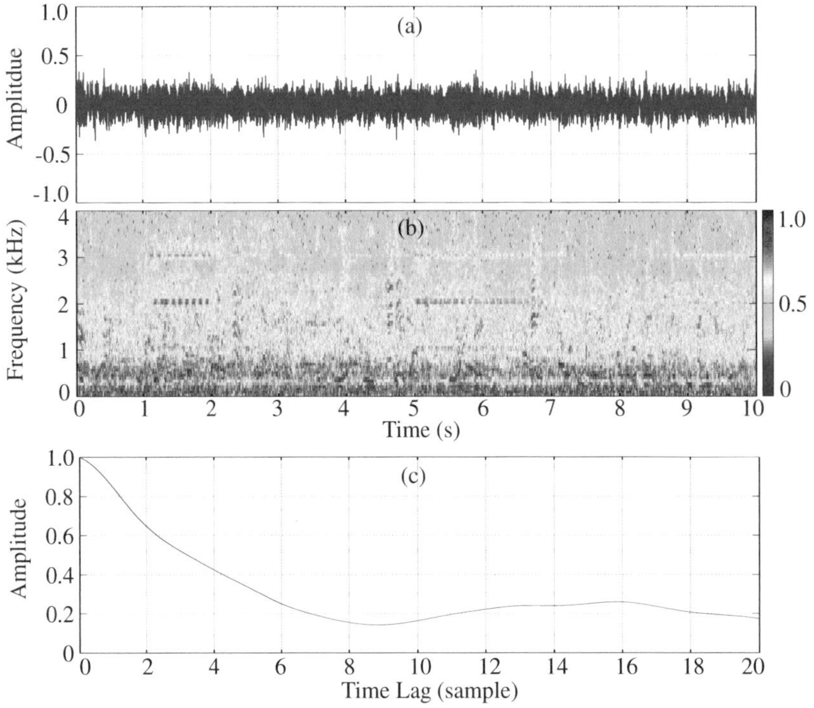

Figure 9.3: The NYSE noise used in the experiments: (a) the first 10-second waveform, (b) the first 10-second spectrogram, and (c) the first 20 autocorrelation coefficients.

are constructed from the corresponding filtered KLE coefficients using the synthesis transform given in (2.8).

9.4 PERFORMANCE OF THE TIME-DOMAIN FILTERS

In this set of experiments, we study the performance of two important time-domain noise reduction filters: Wiener and tradeoff. As described earlier, we compute the noisy signal and noise correlation matrices directly from the noisy and noise signals using a short-time average. Specifically, at each time instant k, an estimate of $\mathbf{R}_\mathbf{y}$, denoted by $\hat{\mathbf{R}}_\mathbf{y}(k)$, is calculated from $y(k)$ using the most recent 320 samples (a 40-ms window length). The matrix $\mathbf{R}_\mathbf{v}$ is computed in a similar way. But noise is supposed to be relatively more stationary than speech, so we use 640 samples (an 80-ms window length) to compute the estimate of $\mathbf{R}_\mathbf{v}$. The Wiener and tradeoff filters are then implemented by substituting $\mathbf{R}_\mathbf{y}$ and $\mathbf{R}_\mathbf{v}$ with $\hat{\mathbf{R}}_\mathbf{y}(k)$ and $\hat{\mathbf{R}}_\mathbf{v}(k)$ into (3.23) and (3.47), respectively. The input SNR for this experiment is set to 10 dB.

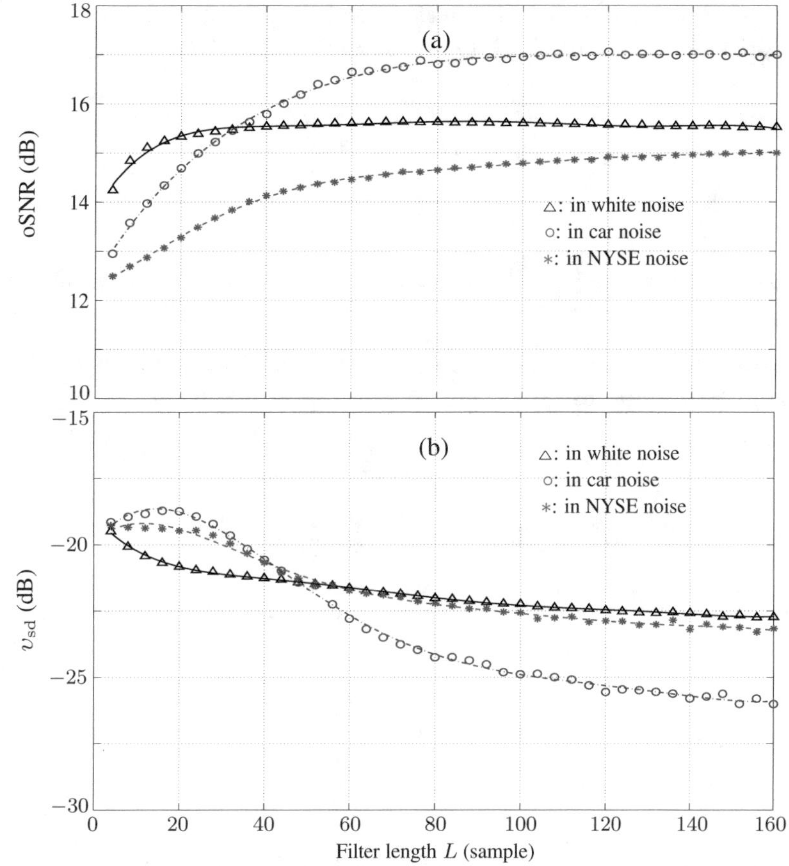

Figure 9.4: Performance of the time-domain Wiener filter as a function of the filter length L: (a) output SNR and (b) speech-distortion index. The input SNR is 10 dB.

9.4.1 WIENER FILTER

With the above experimental setup, the key parameter that affects the performance of the Wiener filter is the filter length L. The optimal value of L depends on many factors such as the degree of autocorrelation of the desired speech signal and that of the noise. Figure 9.4 plots the performance of the Wiener filter as a function of L in three noise conditions (white Gaussian, car, and NYSE). Note that only the output SNR and speech-distortion index are plotted while the noise-reduction factor is omitted in the figure because its curve is similar to that of the output SNR and does not provide much additional information.

In the white Gaussian noise condition, it is seen that the output SNR increases while the speech-distortion index decreases with L at first. But when the value of L is larger than 20, the two measures do not change much with L. The reason can be explained as follows. When the noise is white Gaussian, the product matrix $\mathbf{Q}^T \mathbf{R_v} \mathbf{Q}$ becomes $\sigma_v^2 \mathbf{I}$ and the Wiener filter given in (3.23) can be written as

$$\mathbf{H}_{\mathrm{t,W}} = \mathbf{Q}\boldsymbol{\Sigma}_{\mathrm{t,W}}\mathbf{Q}^T, \tag{9.1}$$

where

$$\boldsymbol{\Sigma}_{\mathrm{t,W}} = \mathrm{diag}\left(\frac{\lambda_{x,1}}{\lambda_{x,1}+\sigma_v^2}, \frac{\lambda_{x,2}}{\lambda_{x,2}+\sigma_v^2}, \ldots, \frac{\lambda_{x,L}}{\lambda_{x,L}+\sigma_v^2}\right) \tag{9.2}$$

is a diagonal matrix, and $\lambda_{x,l}$, $l = 1, 2, \cdots, L$, are the eigenvalues of the matrix $\mathbf{R_x}$ with $\lambda_{x,1} \geq \lambda_{x,2} \geq \ldots \geq \lambda_{x,L} \geq 0$. Substituting (9.1) into (3.7) and (3.9), we obtain

$$\mathrm{oSNR}(\mathbf{H}_{\mathrm{t,W}}) = \frac{\displaystyle\sum_{l=1}^{L} \frac{\lambda_{x,l}^3}{\left(\lambda_{x,l}+\sigma_v^2\right)^2}}{\displaystyle\sum_{l=1}^{L} \frac{\lambda_{x,l}^2 \cdot \sigma_v^2}{\left(\lambda_{x,l}+\sigma_v^2\right)^2}}, \tag{9.3}$$

$$\upsilon_{\mathrm{sd}}\left(\mathbf{H}_{\mathrm{t,W}}\right) = \frac{\displaystyle\sum_{l=1}^{L} \frac{\lambda_{x,l} \cdot \sigma_v^4}{\left(\lambda_{x,l}+\sigma_v^2\right)^2}}{\displaystyle\sum_{l=1}^{L} \lambda_{x,l}}. \tag{9.4}$$

As discussed in Chapter 3, a speech signal is predictable in nature and can be modelled as a linear combination of a small number of (linearly independent) basis vectors. So, the positive semi-definite matrix $\mathbf{R_x}$ has only a limited number of positive eigenvalues and the rest is zero. Let us assume that the number of positive eigenvalues is L_{s}. It is easy to check from (9.3) and (9.4) that once the value of L is greater than L_{s}, further increasing L has no effect on either the output SNR or the speech-distortion index. Of course, the value of L_{s} varies depending on the nature of the sounds in the speech signal. It is relatively small for voiced sounds while large for unvoiced sounds. But in average, the value of L_{s} is around 20 for an 8-kHz sampling rate [11]. That is why in Fig. 9.4, the optimal performance of the Wiener filter in the white Gaussian noise condition is achieved when the value of L is around 20 and further increasing L does not lead to much performance improvement.

If noise is colored, the product matrix $\mathbf{Q}^T \mathbf{R_v} \mathbf{Q}$ is no longer diagonal. In this situation, the Wiener filtering matrix depends not only on the degree of the autocorrelation of the speech signal, but also on that of the noise signal. So, a larger filter length should be used in colored noise as compared to the white noise condition. From Fig. 9.4, it is seen that in both the car and NYSE noise conditions the output SNR increases with L, but it increases more quickly for $L \leq 80$ and for $L > 80$, the improvement in the output SNR is almost negligible. Unlike the output SNR, the speech-distortion index in the car and NYSE noise conditions is not a monotonic function of L. It first increases slightly and then decreases as L increases. This, again, indicates that a larger filter length is needed if the noise is colored. We see that in both conditions, there is not much change in the speech-distortion index for $L > 80$. Therefore, 80 should be a sufficient value of L in the car and NYSE noise conditions.

It is also seen that when the filter length is sufficiently large (e.g., $L > 40$), the Wiener filter achieves the best performance in the car noise condition. Comparatively, the performance in the NYSE noise condition is relatively poorer, which is intuitively reasonable since babble noise is nonstationary and therefore more difficult to deal with than stationary noise.

9.4.2 TRADEOFF FILTER

In the tradeoff filter given in (3.47), a parameter μ is introduced to control the compromise between the amount of noise reduction and the degree of speech distortion. This experiment illustrates the impact of the value of μ on the noise reduction performance. Based on the previous experiment, we set the filter length L to 60, and the results are shown in Fig. 9.5.

When $\mu = 0$, the tradeoff filter becomes the identity matrix, which passes the noisy speech without modifying it. So, there will be neither speech distortion nor noise reduction, which can be seen from Fig. 9.5 where when $\mu = 0$, the output SNR is the same as the input SNR and the speech-distortion index is very small (note that the value of the speech distortion index is less than -100 dB when $\mu = 0$, which is not displayed in the figure). When $\mu = 1$, the tradeoff filter becomes the Wiener one. As we increase the value of μ, a higher output SNR is obtained, but at a price of adding more speech distortion as seen from Fig. 9.5 that both the output SNR and the speech-distortion index increase with μ.

Before leaving this subsection, we want to bring the reader's attention to a numerical issue in implementing the tradeoff filter. From (3.47), one can see that we need to compute the inverse of the sum matrix $\mathbf{R_y} + (\mu - 1)\mathbf{R_v}$. In practice, both matrices $\mathbf{R_y}$ and $\mathbf{R_v}$ are generally positive definite. However, when $\mu < 1$, we subtract a scaled version of $\mathbf{R_v}$ from $\mathbf{R_y}$ and the resulting matrix can become singular. This problem becomes more and more serious when μ decreases from 1 to 0. In the extreme case where $\mu = 0$, we have $\mathbf{R_y} + (\mu - 1)\mathbf{R_v} = \mathbf{R_x}$. As we discussed earlier, when the value of L is large, $\mathbf{R_x}$ is rank deficient, so its inverse does not exist. A straightforward way to circumvent this issue is through the use of a pseudo-inverse when $\mu < 1$, which was adopted in our implementation.

9.5 PERFORMANCE OF THE KLE-DOMAIN FILTERS WITH MODEL 1

In this set of experiments, we study the performance of the KLE domain filters with Model 1. Again, we choose to illustrate the Wiener and tradeoff filters. The matrices $\mathbf{R_y}$ and $\mathbf{R_v}$ are estimated in the same way as in the previous experiments, and the input SNR is set to 10 dB.

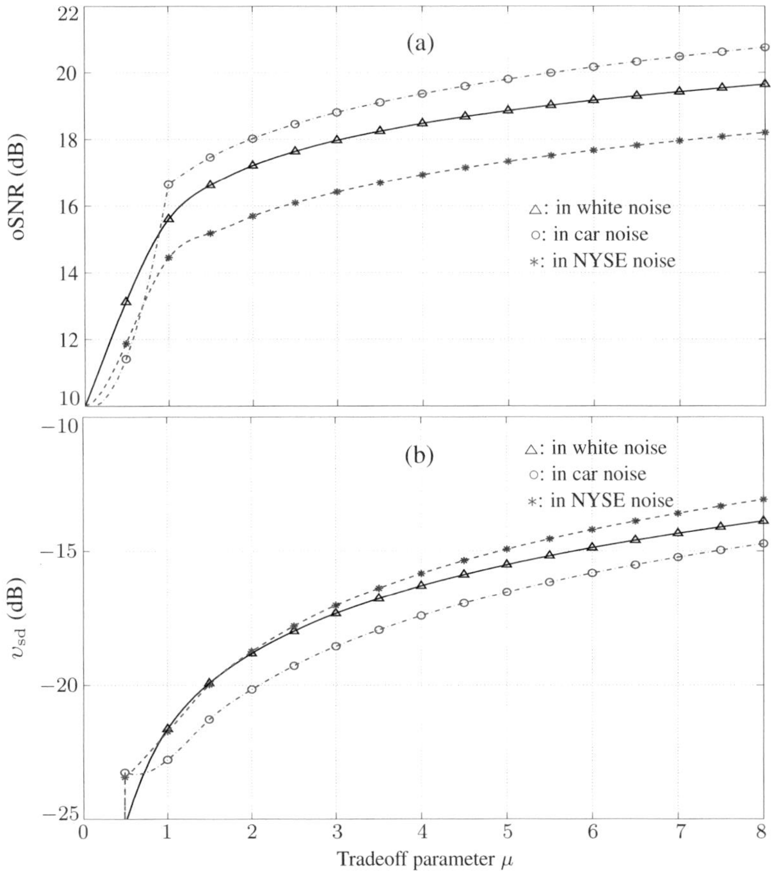

Figure 9.5: Performance of the time-domain tradeoff filter as a function of the parameter μ: (a) output SNR and (b) speech-distortion index. The input SNR is 10 dB and $L = 60$.

9.5.1 KLE-DOMAIN WIENER FILTER

The Wiener filter is implemented according to (5.43) by substituting the matrices $\mathbf{R_y}$ and $\mathbf{R_v}$ with the corresponding estimates. The performance of this Wiener filter as a function of L is sketched in Fig. 9.6. Comparing Figs. 9.6 and 9.4, one can see that relationship between the noise reduction performance and the filter length L of this Wiener filter is similar to that of the time-domain Wiener filter. We notice that the time- and KLE-domain Wiener filters have the same performance in the white Gaussian noise condition. This is due to the fact that the product matrix $\mathbf{Q}^T\mathbf{R_v}\mathbf{Q}$ becomes a diagonal one and, therefore, the two filters are identical. However, in the car and NYSE noise conditions, the KLE-domain Wiener filter has a slightly lower output SNR and a higher speech-

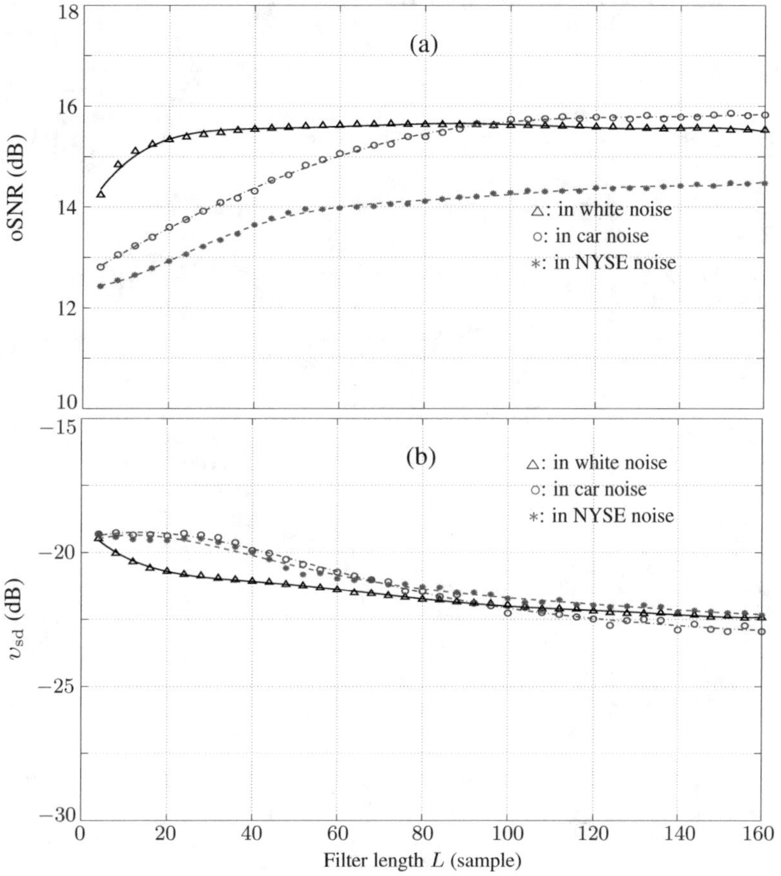

Figure 9.6: Performance of the KLE-domain Wiener filter with Model 1 as a function of the filter length L: (a) output SNR and (b) speech distortion index. The input SNR is 10 dB.

distortion index. The underlying reason will be explained when we discuss the experiments for filters with Model 3.

9.5.2 KLE-DOMAIN TRADEOFF FILTER

The tradeoff filter with Model 1 can be implemented either according to (5.47) or based on (5.53). Here, we choose to use (5.53). Figure 9.7 plots the output SNR and speech-distortion index, both as a function of μ. Comparing Fig. 9.7 with Fig. 9.5, one can see that KLE- and time-domain tradeoff filters have the same performance in white Gaussian noise. This is because the two filters are identical in this condition. However, the performance of the KLE-domain tradeoff filter is inferior to that

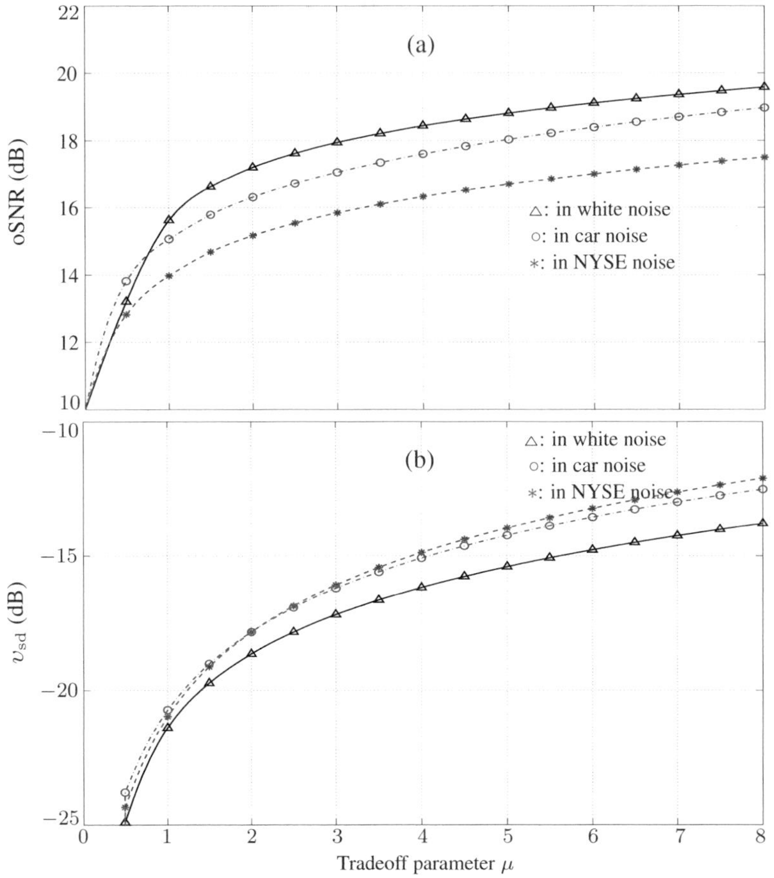

Figure 9.7: Performance of the KLE-domain tradeoff filter with Model 1 as a function of the parameter μ: (a) output SNR and (b) speech-distortion index. The input SNR is 10 dB and $L = 60$.

of its time-domain counterpart for the same value of μ for the other noises. The reason will be explained in the next subsection when we discuss the filters with Model 3.

Before discussing the filters with Model 3, we want to point out that the KLE-domain filters with Model 1 are easier and more efficient to implement as compared to the time-domain filters since the matrix that needs to be inverted in Model 1 is diagonal.

9.6 PERFORMANCE OF THE KLE-DOMAIN FILTERS WITH MODEL 3

Before talking about the performance of the filters with Model 2, let us first discuss the experiments for the filters with Model 3. One may ask why we need the filters with Model 3, given that the KL transform diagonalizes the noisy covariance matrix and, therefore, the KLE coefficients in different subbands should be uncorrelated. The reasons for using the interband information are multiple. In order to estimate the KL transform, we need to obtain an estimate of the noisy covariance matrix $\mathbf{R_y}$. Typically, such an estimate is computed from the noisy signal by using a short-time average [or a recursive method [13]] to approximate the expectation operation. The window length (number of samples) used in the short-time average plays a vital role in the accuracy of the estimated covariance matrix. If the window length is too short, the estimation variance would be very large, which will eventually be translated into less SNR improvement and more speech distortion. In addition, the covariance matrix estimate may not be full rank. To reduce the variance of the $\mathbf{R_y}$ estimate and make it invertible, we need to use a large window length. But with a large window length, the covariance estimate may not be able to follow the time-varying statistics of the speech. As a result, the KL transform that diagonalizes the estimate of the covariance matrix may not diagonalize the true covariance matrix. One easy way to verify this is through examining the cross-correlation between KLE coefficients from different subbands, which will be left to the reader's investigation. Here, we take a different approach to illustrate the existence of interband correlation: we compare the noise reduction performance between Wiener filters with Model 1 and Model 3 by varying the window length N. The results in the car noise condition are plotted in Fig. 9.8. It is seen that the output SNR of the Wiener filter with Model 1 decreases as the window length N increases. If a long-time average is used to estimate the noisy covariance matrix, the output SNR is only 1 dB higher than the input SNR. In comparison, the output SNR for the Wiener filter with Model 3 does not decrease much with N. One can easily notice that the difference between Wiener filters with Model 3 and Model 1 in the output SNR, increases with the window length N. We also see that the Wiener filter with Model 3 has a smaller speech-distortion index. All these points indicate that there exists cross-correlation between the KLE coefficients from different subbands and the degree of this correlation increases as we use a longer window in the short-time average to estimate the noisy covariance matrix. Therefore, it is necessary to use the interband information in developing noise reduction filters, particularly when we use a long window in the short-time average.

While it diagonalizes the noisy covariance matrix, the KL transform may not diagonalize the noise and speech covariance matrices. This is another reason to use the interband information. The only exceptional case is when noise is white. In this situation, the noise covariance is a diagonal matrix, and the KL transform would simultaneously diagonalize both the speech and noisy covariance matrices.

In Chapter 7, we have shown that the KLE-domain noise reduction filters with Model 3 are identical to their counterparts in the time domain. This is intuitively obvious since the time-domain filters are derived using the fullband signal, which is equivalent to using all the self- and cross-band

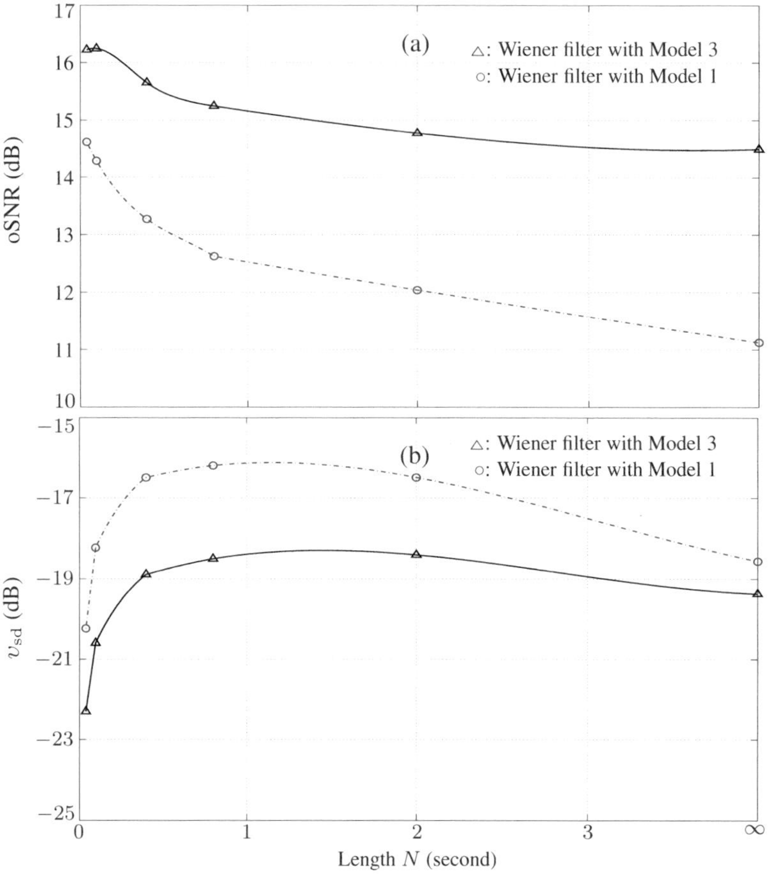

Figure 9.8: Performance of the Wiener filters with Model 1 and Model 3 as a function of the window length N in the car noise condition: (a) output SNR and (b) speech-distortion index. The input SNR is 10 dB and $L = 60$. In the "∞" case, the window length N is equal to the overall length of the $y(k)$ signal. So, the short-time average becomes a long-time average in this situation.

information in the KLE domain. The equivalence between the time-domain filters and the KLE-domain filters with Model 3, on the one hand, naturally explains why the time-domain Wiener and tradeoff filters have better performance than their counterparts in the KLE-domain with Model 1 in colored noise, and on the other hand, demonstrates the motivation for developing filters with Model 3. But filters with Model 1 are simple, may be good enough in most applications, and are equivalent to the optimal gains in the frequency domain.

9.7 PERFORMANCE OF THE KLE-DOMAIN FILTERS WITH MODEL 2

In the previous experiments, we applied a short-time average to estimate the noisy covariance matrix. The window length in the short-time average should be properly selected so that the covariance matrix estimate can follow the nonstationarity of the desired speech signal to achieve noise reduction. Alternatively, the nonstationarity of speech can be employed in the KLE domain. This has led to the development of the filters with Model 2. Unlike Model 1 and Model 3 where each frame may have a different transformation matrix $\mathbf{Q}$, algorithms with Model 2 assume that all the frames share the same matrix $\mathbf{Q}$; otherwise, filtering the KLE coefficients across different frames would not make much sense. With this requirement, the estimation of $\mathbf{Q}$ should be relatively easy: we can simply use a long-term sample average to compute the correlation matrix $\mathbf{R}_{\mathbf{y}}$, and the KL transform matrix $\mathbf{Q}$ can then be obtained by performing the eigenvalue decomposition of $\mathbf{R}_{\mathbf{y}}$. In the course of our study, we found that the estimation accuracy of the matrix $\mathbf{Q}$ plays a less important role in noise reduction performance for the filters with Model 2 than it does for the filters with Model 1 and Model 3. We can even replace the matrix $\mathbf{Q}$ with either the Fourier matrix $\mathbf{F}$ used in the DFT, or the coefficient matrix in the discrete cosine transform (DCT) without degrading noise reduction performance of the filters with Model 2. This is to say that the idea of the filters with Model 2 can also be applied to the frequency-domain approaches. However, strictly following the theoretical development in Chapter 6, we still use the transformation matrix $\mathbf{Q}$ in our experiments with the correlation matrix $\mathbf{R}_{\mathbf{y}}$ being estimated using a long-term average. This matrix $\mathbf{Q}$ is then applied to each frame of the noisy and noise signals to compute the KLE coefficients $c_{y,l}(m)$ and $c_{v,l}(m)$.

The construction of the filters with Model 2 requires to know the subband correlation matrices $\mathbf{\Phi}_{\mathbf{c}_{y,l}}$ and $\mathbf{\Phi}_{\mathbf{c}_{v,l}}$. Again, we can use a short-time average to approximate the mathematical expectation to compute these two matrices. But we found that it is easier to optimize the performance if we use a recursive method as in [13] to estimate $\mathbf{\Phi}_{\mathbf{c}_{y,l}}$ and $\mathbf{\Phi}_{\mathbf{c}_{v,l}}$. Specifically, in this experiment, an estimate of the $\mathbf{\Phi}_{\mathbf{c}_{y,l}}$ matrix at the mth frame is computed using the following recursion:

$$\hat{\mathbf{\Phi}}_{\mathbf{c}_{y,l}}(m) = \alpha_{c_{y,l}} \hat{\mathbf{\Phi}}_{\mathbf{c}_{y,l}}(m-1) + (1 - \alpha_{c_{y,l}})\mathbf{c}_{y,l}(m)\mathbf{c}_{y,l}^T(m), \tag{9.5}$$

where $\alpha_{c_{y,l}}$ is a forgetting factor that controls the influence of the previous data samples on the current estimate of the noisy correlation matrix. The noise covariance matrix $\mathbf{\Phi}_{\mathbf{c}_{v,l}}$ is estimated in a similar manner but with a different forgetting factor $\alpha_{c_{v,l}}$. With the estimated covariance matrices $\hat{\mathbf{\Phi}}_{\mathbf{c}_{y,l}}(m)$ and $\hat{\mathbf{\Phi}}_{\mathbf{c}_{v,l}}(m)$, an estimate of the $\mathbf{\Phi}_{\mathbf{c}_{x,l}}$ matrix at time m is computed as $\hat{\mathbf{\Phi}}_{\mathbf{c}_{x,l}}(m) = \hat{\mathbf{\Phi}}_{\mathbf{c}_{y,l}}(m) - \hat{\mathbf{\Phi}}_{\mathbf{c}_{v,l}}(m)$ and the interframe correlation vector at time m, i.e., $\hat{\boldsymbol{\gamma}}_{\mathbf{c}_{x,l}}(m)$, is taken as the first column of $\hat{\mathbf{\Phi}}_{\mathbf{c}_{x,l}}(m)$ normalized with its first element.

Substituting $\hat{\mathbf{\Phi}}_{\mathbf{c}_{y,l}}(m)$, $\hat{\mathbf{\Phi}}_{\mathbf{c}_{v,l}}(m)$ and $\hat{\boldsymbol{\gamma}}_{\mathbf{c}_{x,l}}(m)$ into (6.40) and (6.56), we implemented the Wiener and MVDR filters with Model 2. There are many parameters that affect the performance of the filters with Model 2, such as the transformation length L, the filter length M, and the forgetting factors $\alpha_{c_{y,l}}$ and $\alpha_{c_{v,l}}$. These parameters can be tuned up through experiments step by step by varying

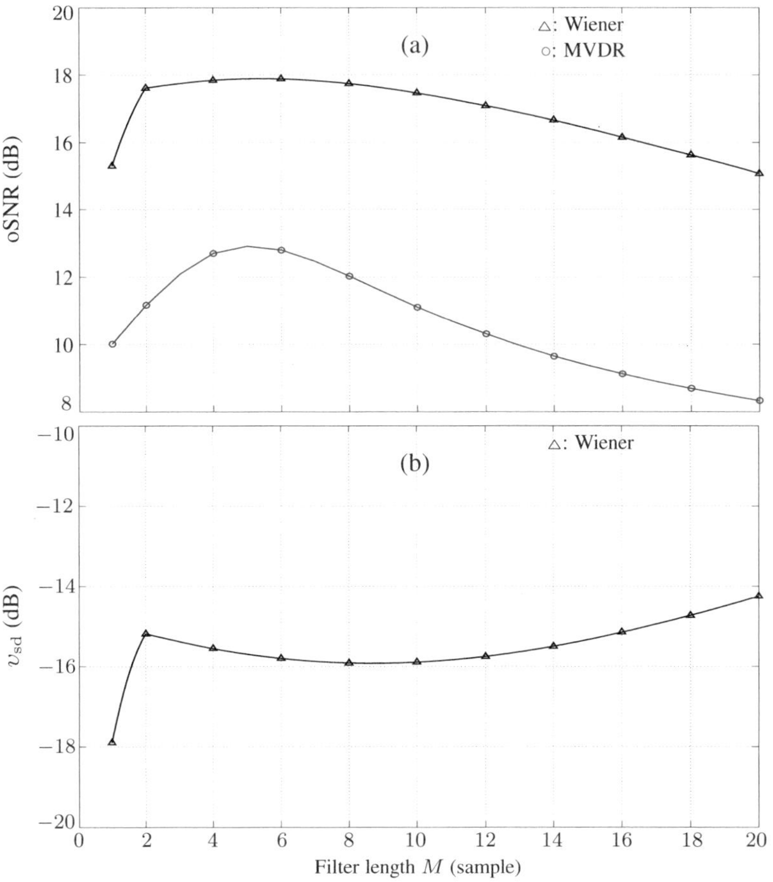

Figure 9.9: Performance of the Wiener and MVDR filters with Model 2 as a function of the filter length M in the white Gaussian noise condition: (a) output SNR and (b) speech-distortion index. The input SNR is 10 dB and $L = 20$. Note that the speech-distortion index for the MVDR filter is smaller than -100 dB and it is not displayed.

one parameter while fixing the others at a time. Let us set $L, \alpha_{c_{y,l}}$, and $\alpha_{c_{v,l}}$, respectively, to 20, 0.8, and 0.9 following the results in [13] and study the impact of the filter length M on the output SNR and speech-distortion index. The results for both the Wiener and MVDR filters in the white Gaussian noise condition are shown in Fig. 9.9.

It is seen from Fig. 9.9 that the output SNR for both the Wiener and MVDR filters first increases and then decreases as the filter length M increases. With a properly selected value of M, one can see that a better performance is achieved with the Model 2 Wiener filter as compared to the Model 1 Wiener filter, which justifies the motivation of using interframe information. We also see that for $M > 8$, if we keep increasing M, there is some performance degradation. The reason can probably be explained as follows. With the same forgetting factor, the matrix $\hat{\mathbf{\Phi}}_{\mathbf{c}_{y,l}}(m)$ would be less well conditioned for a larger value of M, and inverting this matrix may cause some numerical problems. One way to overcome this problem is to use either a larger forgetting factor or using more regularization when computing the inverse of $\hat{\mathbf{\Phi}}_{\mathbf{c}_{y,l}}(m)$. We will leave this to the reader's own exploration.

It is interesting to see that with the Model 2, we can derive an MVDR filter. When $M = 1$, there will be neither noise reduction nor speech distortion. But as we increase M to larger than 1, we can achieve some noise reduction without adding speech distortion. Note that the speech distortion index is smaller than -100 dB for the MVDR filter, which is not displayed in Fig. 9.9.

We also notice some difference between the results in Fig. 9.9 and those in [13]. This is because the whole filtered speech is treated as the desired speech component in [13] while in this book we have divided the filtered speech into a desired speech and an interference component. Apparently, treating the interference component as noise after noise reduction is more reasonable since it is uncorrelated with the desired speech samples that we want to estimate.

9.8 KLE-DOMAIN FILTERS WITH MODEL 4

Similar to the filters with Model 2, the filters with Model 4 requires to have the same transformation matrix $\mathbf{Q}$ across different frames. With this framework, we can use a long-term average to estimate the correlation matrix $\mathbf{R}_{\mathbf{y}}$, thereby obtaining the matrix $\mathbf{Q}$. We can also replace the matrix $\mathbf{Q}$ with either the Fourier matrix $\mathbf{F}$ or the DCT matrix.

From the previous experiments with Model 2, we already see that taking into account the interframe information can help improve noise reduction performance. We have also demonstrated in the Model 3 that the interband information is needed to better cope with colored noise. This justifies the motivation of developing the filters with Model 4. But we will leave the performance study of the filters with Model 4 to the reader's own investigation.

Bibliography

[1] J. K. Baker, "The dragon system–An overview," *IEEE Trans. Acoust., Speech, Signal Process.*, vol. ASSP-23, pp. 24–29, Feb. 1975. DOI: 10.1109/TASSP.1975.1162650 3

[2] L. E. Baum and T. Petrie, "Statisitcal inference for probabilistic functons of finite state Markov chains," *Ann. Math. Stat.*, vol. 73, pp. 1554–1563, 1966. DOI: 10.1214/aoms/1177699147 3

[3] J. Benesty and T. Gaensler, "New insights into the RLS algorithm," *EURASIP J. Applied Signal Process.*, vol. 2004, pp. 331–339, Mar. 2004. DOI: 10.1155/S1110865704310188 17

[4] J. Benesty, S. Makino, and J. Chen, Eds., *Speech Enhancement*. Berlin, Germany: Springer-Verlag, 2005. 4, 11

[5] J. Benesty, J. Chen, Y. Huang, and S. Doclo, "Study of the Wiener filter for noise reduction," in *Speech Enhancement*, J. Benesty, S. Makino, and J. Chen, Eds., Berlin, Germany: Springer-Verlag, 2005, Chapter 2, pp. 9–41. 11, 12, 13

[6] J. Benesty, J. Chen, and Y. Huang, *Microphone Array Signal Processing*. Berlin, Germany: Springer-Verlag, 2008. 1

[7] J. Benesty, J. Chen, Y. Huang, and I. Cohen, *Noise Reduction in Speech Processing*. Berlin, Germany: Springer-Verlag, 2009. 4, 5, 7, 8, 10, 13, 16, 19, 25, 33, 34, 38, 39, 40, 44, 50

[8] J. Benesty, J. Chen, and Y. Huang, "On noise reduction in the Karhunen-Loeve expansion domain," in *Proc. IEEE ICASSP*, 2009, pp. 25–28. DOI: 10.1109/ICASSP.2009.4959511 5, 8, 10

[9] S. F. Boll, "Suppression of acoustic noise in speech using spectral subtraction," *IEEE Trans. Acoust., Speech, Signal Process.*, vol. ASSP-27, pp. 113–120, Apr. 1979. DOI: 10.1109/TASSP.1979.1163209 3

[10] J. Capon, "High resolution frequency-wavenumber spectrum analysis," *Proc. IEEE*, vol. 57, pp. 1408–1418, Aug. 1969. DOI: 10.1109/PROC.1969.7278 50

[11] J. Chen, J. Benesty, Y. Huang, and S. Doclo, "New insights into the noise reduction Wiener filter," *IEEE Trans. Audio, Speech, Language Process.*, vol. 14, pp. 1218–1234, July 2006. DOI: 10.1109/TSA.2005.860851 4, 12, 13, 80

[12] J. Chen, J. Benesty, Y. Huang, and E. J. Diethorn, "Fundamentals of noise reduction," in *Springer Handbook of Speech Processing*, J. Benesty, M. M. Sondhi, and Y. Huang, Eds., Berlin, Germany: Springer-Verlag, 2007, Chapter 43, Part H, pp. 843–872. 11

[13] J. Chen, J. Benesty, and Y. Huang, "Study of the noise-reduction problem in the Karhunen-Loeve expansion domain," *IEEE Trans. Audio, Speech, Language Process.*, vol. 17, pp. 787–802, May 2009. DOI: 10.1109/TASL.2009.2014793 5, 8, 10, 85, 87, 88, 89

[14] M. Dendrinos, S. Bakamidis, and G. Carayannis, "Speech enhancement from noise: a regenerative approach," *Speech Commun.*, vol. 10, pp. 45–57, Feb. 1991. DOI: 10.1016/0167-6393(91)90027-Q 5, 20

[15] Y. Ephraim and D. Malah, "Speech enhancement using a minimum mean-square error short-time spectral amplitude estimator," *IEEE Trans. Acoust., Speech, Signal Process.*, vol. ASSP-32, pp. 1109–1121, Dec. 1984. DOI: 10.1109/TASSP.1984.1164453 3

[16] Y. Ephraim and D. Malah, "Speech enhancement using a minimum mean-square error log-spectral amplitude estimator," *IEEE Trans. Acoust., Speech, Signal Process.*, vol. ASSP-33, pp. 443–445, Apr. 1985. DOI: 10.1109/TASSP.1985.1164550 3

[17] Y. Ephraim, D. Malah, and B.-H. Juang, "On the application of hidden Markov models for enhancing noisy speech," *IEEE Trans. Acoust., Speech, Signal Process.*, vol. ASSP-37, pp. 1846–1856, Dec. 1989. DOI: 10.1109/29.45532 3

[18] Y. Ephraim, "A Bayesian estimation approach for speech enhancement using hidden Markov models," *IEEE Trans. Signal Process.*, vol. 40, pp. 725–735, Apr. 1992. DOI: 10.1109/78.127947 3

[19] Y. Ephraim, "Statstical-model-based speech enhancement systems," *Proc. IEEE*, vol. 80, pp. 1526–1555, Oct. 1992. DOI: 10.1109/5.168664 3

[20] Y. Ephraim and H. L. Van Trees, "A signal subspace approach for speech enhancement," *IEEE Trans. Speech, Audio Process.*, vol. 3, pp. 251–266, July 1995. DOI: 10.1109/89.397090 5, 14, 20

[21] K. Fukunaga, *Introduction to Statistical Pattern Recognition.* San Diego, CA: Academic Press, 1990. 20

[22] S. Gannot, D. Burshtein, and E. Weinstein, "Iterative and sequential Kalman filter-based speech enhancement algorithms," *IEEE Trans. Speech, Audio Process.*, vol. 6, pp. 373–385, July 1998. DOI: 10.1109/89.701367 4

[23] Z. Goh, K.-C. Tan, and B. T. G. Tan, "Kalman-filtering speech enhancement method based on a voiced-unvoiced speech model," *IEEE Trans. Speech, Audio Process.*, vol. 7, pp. 510–524, Sept. 1999. DOI: 10.1109/89.784103 4

[24] G. H. Golub and C. F. Van Loan, *Matrix Computations*. Baltimore, MD: The Johns Hopkins University Press, 1996. 8

[25] S. Haykin, *Adaptive Filter Theory*. Fourth Edition, Upper Saddle River, NJ: Prentice-Hall, 2002. 8

[26] K. Hermus, P. Wambacq, and H. Van hamme, "A review of signal subspace speech enhancement and its application to noise robust speech recognition," *EURASIP J. Advances Signal Process.*, vol. 2007, Article ID 45821, 15 pages, 2007. DOI: 10.1155/2007/45821 20

[27] Y. Hu and P. C. Loizou, "A subspace approach for enhancing speech corrupted by colored noise," *IEEE Signal Process. Lett.*, vol. 9, pp. 204–206, July 2002. DOI: 10.1109/LSP.2002.801721 5, 20

[28] Y. Hu and P. C. Loizou, "A subspace approach for enhancing speech corrupted by colored noise," in *Proc. IEEE ICASSP*, 2002, pp. I-573–I-576. DOI: 10.1109/ICASSP.2002.1005804 5, 20

[29] Y. Hu and P. C. Loizou, "A generalized subspace approach for enhancing speech corrupted by colored noise," *IEEE Trans. Speech Audio Process.*, vol. 11, pp. 334–341, July 2003. DOI: 10.1109/TSA.2003.814458 5, 20

[30] Y. Huang, J. Benesty, and J. Chen, *Acoustic MIMO Signal Processing*. Berlin, Germany: Springer-Verlag, 2006. 1, 11

[31] F. Jabloun and B. Champagne, "Signal subspace techniques for speech enhancement," in *Speech Enhancement*, J. Benesty, S. Makino, and J. Chen, Eds., Berlin, Germany: Springer-Verlag, 2005, Chapter 7, pp. 135–159. 20

[32] F. Jelinek, "Continuous speech recognitiion by statistical methods," *Proc. IEEE*, vol. 64, pp. 532–536, Apr. 1976. DOI: 10.1109/PROC.1976.10159 3

[33] S. H. Jensen, P. C. Hansen, S. D. Hansen, and J. A. Sorensen, "Reduction of broad-band noise in speech by truncated QSVD," *IEEE Trans. Speech Audio Process.*, vol. 3, pp. 439–448, Nov. 1995. DOI: 10.1109/89.482211 5, 20

[34] S. Kay, "Some results in linear interpolation theory," *IEEE Trans. Acoust., Speech, Signal Process.*, vol. ASSP-31, pp. 746–749, June 1983. DOI: 10.1109/TASSP.1983.1164088 17

[35] B. Koo and J. D. Gibson, "Filtering of colored noise for speech enhancement and coding," in *Proc. IEEE ICASSP*, 1989, pp. 345–352. DOI: 10.1109/78.91144 4

[36] R. T. Lacoss, "Data adaptive spectral analysis methods," *Geophysics*, vol. 36, pp. 661–675, Aug. 1971. DOI: 10.1190/1.1440203 50

[37] B. Lee, K. Y. Lee, and S. Ann, "An EM-based approach for parameter enhancement with an application to speech signals," *Signal Process.*, vol. 46, pp. 1–14, Sept. 1995. DOI: 10.1016/0165-1684(95)00068-O 4

[38] C. Li and S. Vang Andersen, "Inter-frequency dependency in MMSE speech enhancement," in *Proc. NORSIG*, 2004, pp. 200–203. DOI: 10.1109/NORSIG.2004.250161 9

[39] J. S. Lim and A. V. Oppenheim, "Enhancement and bandwidth compression of noisy speech," *Proc. IEEE*, vol. 67, pp. 1586–1604, Dec. 1979. DOI: 10.1109/PROC.1979.11540 3

[40] P. Loizou, *Speech Enhancement: Theory and Practice*. Boca Raton, FL: CRC Press, 2007. 11

[41] R. J. McAulay and M. L. Malpass, "Speech enhancement using a soft-decision noise suppression filter," *IEEE Trans. Acoust., Speech, Signal Process.*, vol. ASSP-28, pp. 137–145, Apr. 1980. DOI: 10.1109/TASSP.1980.1163394 3

[42] M. Niedźwiecki and K. Cisowski, "Adaptive scheme for elimination of broadband noise and impulsive disturbances from AR and ARMA signals," *IEEE Trans. Signal Process.*, vol. 44, pp. 528–537, Mar. 1996. DOI: 10.1109/78.489026 4

[43] K. K. Paliwal and A. Basu, "A speech enhancement method based on Kalman filtering," in *Proc. IEEE ICASSP*, 1987, pp. 177–180. DOI: 10.1109/ICASSP.1987.1169756 4

[44] L. R. Rabiner, "A tutorial on hidden Markov models and selected applications in speech recognition," *Proc. IEEE*, vol. 77, pp. 257–286, Feb. 1989. DOI: 10.1109/5.18626 3

[45] L. R. Rabiner and B. H. Juang, *Fundamentals of Speech Recognition*. Englewood Cliffs, NJ: Prentice-Hall, 1993.

[46] A. Rezayee and S. Gazor, "An adpative KLT approach for speech enhancement," *IEEE Trans. Speech Audio Process.*, vol. 9, pp. 87–95, Feb. 2001. DOI: 10.1109/89.902276 41

[47] H. Sameti, H. Sheikhzadeh, L. Deng, and R. L. Brennan, "HMM-based strategies for enhancement of speech signals embedded in nonstationary noise," *IEEE Trans. Speech, Audio Process.*, vol. 6, pp. 445–455, Sept. 1998. DOI: 10.1109/89.709670 3

[48] M. R. Schroeder, "Apparatus for suppressing noise and distortion in communication signals," U.S. Patent No. 3,180,936, filed Dec. 1, 1960, issued Apr. 27, 1965. 2

[49] M. Souden, J. Benesty, and S. Affes, "On the global output SNR of the parameterized frequency-domain multichannel noise reduction Wiener filter," *IEEE Signal Process. Lett.*, vol. 17, pp. 425-428, May 2010. DOI: 10.1109/LSP.2010.2042520 53

[50] P. Vary and R. Martin, *Digital Speech Transmission: Enhancement, Coding and Error Concealment*. Chichester, England: John Wiley & Sons Ltd, 2006. 7, 11

[51] P. J. Wolfe and S. J. Godsill, "Simple alternatives to the Ephraim and Malah suppression rule for speech ehancemnet," in *Proc. IEEE ICASSP*, 2001, pp. 496–499. DOI: 10.1109/SSP.2001.955331 3

Authors' Biographies

JACOB BENESTY

Jacob Benesty was born in 1963. He received a Masters degree in microwaves from Pierre & Marie Curie University, France, in 1987, and a Ph.D. degree in control and signal processing from Orsay University, France, in April 1991. During his Ph.D. program (from November 1989 to April 1991), he worked on adaptive filters and fast algorithms at the Centre National d'Etudes des Telecommunications (CNET), Paris, France. From January 1994 to July 1995, he worked at Telecom Paris University on multichannel adaptive filters and acoustic echo cancellation. From October 1995 to May 2003, he was first a Consultant and then a Member of the Technical Staff at Bell Laboratories, Murray Hill, NJ, USA. In May 2003, he joined INRS-EMT, University of Quebec, in Montreal, Quebec, Canada, as a Professor. His research interests are in signal processing, acoustic signal processing, and multimedia communications. Dr. Benesty received the 2001 and 2008 Best Paper Awards from the IEEE Signal Processing Society. In 2010, he received the Gheorghe Cartianu Award from the Romanian Academy. He was a member of the editorial board of the EURASIP Journal on Applied Signal Processing, a member of the IEEE Audio & Electroacoustics Technical Committee, the co-chair of the 1999 International Workshop on Acoustic Echo and Noise Control (IWAENC), and the general co-chair of the 2009 IEEE Workshop on Applications of Signal Processing to Audio and Acoustics (WASPAA). Dr. Benesty co-authored and co-edited many books in the area of acoustic signal processing.

JINGDONG CHEN

Jingdong Chen received B.S. and M.S. degrees in electrical engineering from the Northwestern Polytechnic University, Xiaan, China, in 1993 and 1995, respectively, and the Ph.D. degree in pattern recognition and intelligence control from the Chinese Academy of Sciences, Beijing, in 1998. From 1998 to 1999, he was with ATR Interpreting Telecommunications Research Laboratories, Kyoto, Japan, where he conducted research on speech synthesis, speech analysis, as well as objective measurements for evaluating speech synthesis. He then joined the Griffith University, Brisbane, Australia, as a Research Fellow, where he engaged in research in robust speech recognition and signal processing. From 2000 to 2001, he worked at ATR Spoken Language Translation Research Laboratories on robust speech recognition and speech enhancement. From 2001 to 2009, he was a Member of Technical Staff at Bell Laboratories, Murray Hill, New Jersey, working on acoustic signal processing for telecommunications. He is currently serving as the Chief Scientist of WeVoice Inc. in New Jersey. His research interests include adaptive signal processing, speech enhancement, adaptive noise/echo cancellation, microphone array signal processing, signal separation, and source localization. Dr. Chen co-authored and co-edited several books in the area of acoustic and speech signal processing. He is currently an Associate Editor of the IEEE Transactions on Audio, Speech, and Language Processing, a member of the IEEE Audio and Electroacoustics Technical Committee, and a member of the editorial board of the Open Signal Processing Journal. He helped organize the 2005 IEEE Workshop on Applications of Signal Processing to Audio and Acoustics (WASPAA), and was the technical Co-Chair of the 2009 WASPAA. Dr. Chen received the 2008 Best Paper Award from the IEEE Signal Processing Society, the 1998-1999 Japan Trust International Research Grant from the Japan Key Technology Center, the Young Author Best Paper Award from the 5th National Conference on Man-Machine Speech Communications, and the 1998 President's Award from the Chinese Academy of Sciences.

YITENG HUANG

Yiteng Huang received his B.S. degree from the Tsinghua University, Beijing, China, in 1994 and the M.S. and Ph.D. degrees from the Georgia Institute of Technology (Georgia Tech), Atlanta, in 1998 and 2001, respectively, all in electrical and computer engineering. From March 2001 to January 2008, he was a Member of Technical Staff at Bell Laboratories, Murray Hill, NJ. In January 2008, he founded the WeVoice, Inc., in Bridgewater, New Jersey and served as its CTO. His current research interests are in acoustic signal processing, multimedia communications, and wireless sensor networks. Dr. Huang served as an Associate Editor for the EURASIP Journal on Applied Signal Processing from 2004 and 2008 and for the IEEE Signal Processing Letters from 2002 to 2005. He served as a technical Co-Chair of the 2005 Joint Workshop on Hands-Free Speech Communication and Microphone Array and the 2009 IEEE Workshop on Applications of Signal Processing to Audio and Acoustics. He is a coeditor/coauthor of the books Noise Reduction in Speech Processing (Springer-Verlag, 2009), Microphone Array Signal Processing(Springer-Verlag, 2008), Springer Handbook of Speech Processing (Springer-Verlag, 2007), Acoustic MIMO Signal Processing (Springer-Verlag, 2006), Audio Signal Processing for Next-Generation Multimedia Communication Systems (Kluwer, 2004), and Adaptive Signal Processing: Applications to Real-World Problems (Springer-Verlag, 2003). He received the 2008 Best Paper Award and the 2002 Young Author Best Paper Award from the IEEE Signal Processing Society, the 2000-2001 Outstanding Graduate Teaching Assistant Award from the School Electrical and Computer Engineering, Georgia Tech, the 2000 Outstanding Research Award from the Center of Signal and Image Processing, Georgia Tech, and the 1997-1998 Colonel Oscar P. Cleaver Outstanding Graduate Student Award from the School of Electrical and Computer Engineering, Georgia Tech.

Index